Micro and Nanomachining Technology- Size, Model and Complex Mechanism

Authored By

Xuesong Han

School of Mechanical Engineering
Tianjin University
P.R.China

Bentham Science Publishers
Executive Suite Y - 2
PO Box 7917, Saif Zone
Sharjah, U.A.E.
subscriptions@benthamscience.org

Bentham Science Publishers
P.O. Box 446
Oak Park, IL 60301-0446
USA
subscriptions@benthamscience.org

Bentham Science Publishers
P.O. Box 294
1400 AG Bussum
THE NETHERLANDS
subscriptions@benthamscience.org

CONTENTS

FOREWORD

Metal cutting is a subject as old as the Industrial Revolution, but one that has evolved continuously as technology has advanced. Currently, the development of products at the nano- and microscale is being driven by economic necessity in order to improve the quality of life. Although the miniature devices may be manufactured by various procedures, their shaping through removal of material constitutes a major means of production. Established and recently developed methods of machining continue to be investigated for the shaping of such parts to specified small dimensions. The term ''micro and nanomachining'' has thus emerged and is generally used to define the practice of material removal for the production of parts at small scale.

The international technology assessment study has focused on the emerging global trend toward the miniaturization of manufacturing processes, equipment and systems for microscale components and products. This ultraprecision machining (or micro & nanomachining technology-MNT) approaches can begin to be compartmentalized by separation into either a ''top-down'' or ''bottom-up'' approach. Top-down fabrication refers to methods where one begins with a macroscopically dimensioned material while bottom-up fabrication refers to build up structures atom by atom. The study of MNT has investigated both the state-of-the-art as well as emerging technologies from the scientific, technological, and commercialization perspectives. This will undoubtedly have serious long-term implications since it is well-recognized that micromanufacturing will be a critical enabling technology in bridging the gap between nanoscience and technology developments and their realization in useful products and processes. It is assured that the importance and number of MNT will grow dramatically in the coming years with the rise of nanotechnology. It is clear that scientists and engineers working from many different directions and finding their inspiration from engineering, chemical, and physical sources, are all contributing greatly to the field of MNT. Thus, it is our belief that readers from all fields will find material of interest in this multidisciplinary topic, and perhaps even find additional inspiration to invent the next-generation MNT methods and tools.

The eBook edited by Prof. Xuesong Han provides a broad overview of the micro and nanomachining technology and the recent advancement. Chapters written by him in his research fields will make the reader acquainted with a variety of topics ranging from the role of MNT, the basic investigation methods and the meaningful results. As such, the volume should be particularly useful to basic investigators, engineers and postgraduates interested in the latest advances in this exciting field.

Yong X. Gan

Associate Professor
Department of Mechanical Engineering
California State Polytechnic University
Pomona, CA
USA

PREFACE

Nanotechnology is seen as the next step in the industrial revolution and, as such, requires machining processes that will revolutionize the way small products are made. It is clear that micro & nanomachining has already been a daunting task. A remarkable enhancement in computational capability (computer hardware) and high performance computation techniques (parallel computation) has enable us to employ the large scale parallel numerical simulation to investigate the micro & nanomachining and gain insights into this process. Process modeling and optimization with the help of computers can reduce expensive and time consuming experiments for manufacturing good quality products. Micro & nanomachining is not the similarity reduction but should be some new mechanism plays a dominant role. This eBook aims to provide the fundamentals and the recent advances in micro and nanomachining for modern manufacturing engineering.

The eBook is divided into five chapters. Chapter 1 is an introduction to micro & nano machining technology (MNT). A brief history of MNT is provided here to highlight the important roles this field has played in manufacturing technology development. The historical development of different MNT is introduced in the second half of this chapter. Further advancements in the MNT can be aided through a theoretical understanding of micro & nanomachining. The diversity of these processes renders difficult a generalized theoretical analysis of MNT; however, the methods of molecular dynamics are becoming increasingly attractive for studies of MNT, especially as the technology advances toward the shaping of parts in the nanometric range. The foundations of molecular dynamics that are needed for theoretical treatments of micromachining therefore form the basis of Chapter 2. Chapter 3 focuses on the fundamental principles of finite element as well as some key factors on numerical modeling of micro & nanomachining and current advancements. Chapter 4 discusses the recent advances on multiscale method, its theoretical basis and their roles on investigation of MNT. Owing to the combination of constantly increasing computational power and the increased knowledge and understanding of material behavior, multiple scale modeling methods have recently emerged as the tool of choice to link the mechanical behavior of materials from the smallest scale of atoms to the largest scale of structures. Chapter 5 focuses on the complexity of MNT. The self-organization of MNT, energy dissipation, and fractal analysis of machined surface acquired by MNT are given in this chapter.

The author thanks a large number of groups and people who have directly or indirectly contributed to making this eBook possible. Many thanks are due and sincerely given to Prof. Bin Lin for putting part of the cooperation work into this eBook. I also wish to note my appreciation to Mr. Xiaohu Liang for helping me diligently in providing corresponding materials.

Most of all, I would like to thank my family for the understanding and love that made writing of this eBook possible.

ACKNOWLEDGEMENT

This work is supported by China Scholarship Council (Grant No. 201208120091).

CONFLICT OF INTEREST

The author confirms that this eBook content has no conflict of interest.

Xuesong Han

School of Mechanical Engineering
Tianjin University
P.R.China
E-mail: hanxuesongphd@gmail.com

List of Contributors

Xuesong Han: Department of Mechanical Engineering and School of Mechanical Enginering, Tianjin University, Tianjin, China

Xiaohu Liang: Department of Mechanical Engineering and School of Mechanical Enginering, Tianjin University, Tianjin, China

Bin Lin: Department of Mechanical Engineering and School of Mechanical Enginering, Tianjin University, Tianjin, China

2

Send Orders for Reprints to reprints@benthamscience.net
Micro and Nanomachining Technology Size, Model and Complex Mechanism, 2014, 3-48   **3**

CHAPTER 1

Introduction to Micro and Nanomachining Technology

Abstract: This chapter focused on the emerging global trend toward the miniaturization of manufacturing processes, equipment and systems for micro and nanoscale components and products, *i.e.*, small equipment for small parts. The present need for smallness of parts stems mainly from two requirements: greater compactness in the utilization of space and portability. The mechanical and electrical devices that make up these items need to be produced in ever-decreasing sizes, with tightly specified dimensions and accuracies. Although these miniature devices may be machined by various techniques, their shaping through material removal constitutes a major means of production. Innovations in the area of micro and nanofabrication have created opportunity to manufacture structures at the nanometer and millimeter scales. These ultraprecision machining processes include STM based nanofabrication and abrasive machining (including lapping, polishing and honing) which can be characterized by either two body or three body abrasive interactions. There are various ways to classify precision material removal processes. We have presented one above, based on the "uncut chip thickness". Combinations of these techniques and established methods of manufacturing that produce hybrid manufacturing processes will create the short term "stepping stones" required to meet the demand generated to economically manufacture microscale products. In this chapter, some dominant micro and nanomachining techniques that are currently used to fabricate structures in the nanometer scale up to the millimeter scale are introduced.

Keywords: Micro and Nanomachining, subtractive microscale process, additive microscale process, size effect, surface.

1. MICRO AND NANOMACHINING TECHNOLOGY

Nanoscience has been making great strides over the past years with breakthroughs coming at a surprisingly rapid rate. It was envisioned in 1959 by Richard Feynman, the Nobel Laureate in physics, presented his vision on miniaturization at the California Institute of Technology of his famous prediction "There are plenty of rooms at the bottom". Feynman expatiated the scaling down of lathes and drilling machines, and talks about drilling holes, turning, molding, stamping parts, and so forth. Feynman also described the need for micro- and nanomachining as the basis or creating a microscopic world that would benefit mankind. Nanotechnology encompasses technology performed at the nanoscale

Xuesong Han

that has real-world applications. Nanotechnology will have a profound effect on our society that will lead to breakthrough discoveries in materials and manufacturing, electronics, medicine, healthcare, the environment, sustainability, energy, biotechnology, information technology, national security and so on. Examples of these components are optical mirrors, computer memory discs, and drums for photocopying machines, with a surface finish in the nanometer range and surface form accuracy in the micron or the sub-micron range. Micro & Nanomachining is one of the keys to the development of novel materials, devices and systems. Precise control of nanomaterials, nanostructures, nanodevices and their performances is essential for future innovations in technology. Nanomachining includes methods that manipulate atoms and molecules to produce single artifacts to produce submicron-sized components and systems and so it is a challenge presented to us to produce single-nanoscale artifacts in a mass production fashion that obviously produces the accompanying economies of scale. The micro and nanomachining technology (MNT) has already become the key factor to ensure success in the international competition. Many parts with high precision level required to be manufactured using MNT. The advanced apparatus manufactured using MNT is needed to develop sophisticated technique, national defense technology, microelectronics industry and so on. MNT is the frontier of modern manufacturing technology, at the same time also being the foundation of tomorrow technology.

According to McKeown [1], microtechnology means the physical scale of the products is small (in manufacturing terms being made to dimensions and tolerances of the order of micrometers (10^{-6}m) and nanotechnology, in which dimensions and tolerances are of the order of nanometers (10^{-9}m). Characteristic of micro & nanomachining is the volume or size of the part removed from the workpiece, termed the "small unit removal" (SUR). For example, in mechanical operations, the SUR consists of the feed, and depth of cut and length corresponding to one chip of material removed; in electro-discharge machining the SUR is defined as the crater produced by one pulse of discharge. Micromachining with masks can yield unit removal as small as the size of atoms. Nanomachining has been defined as an approach to design, produce, control, modify, manipulate, and assemble nanometer-scale elements or features for the purpose of realizing a product or system that exploits properties seen at the nanoscale. Nanomachining R&D has as its goal enabling the mass production of reliable and economical nanoscale materials, structures, devices, and systems. It includes bottom-up directed assembling of nanostructure building blocks (from the atomic, molecular, supramolecular levels); top-down, high-resolution processing (ultraprecision engineering, fragmentation methods); physical-chemical engineering of molecules and supra-molecular systems (molecules as devices "by design", nanoscale machines, *etc.*); and, hierarchical integration with larger scale systems. This requires a high degree of process control in sensing and actuation of matter at the nanoscale, as well as capabilities for scaling-up. One major goal is minimizing use of materials and energy, reduction of waste and environmental impact, and enabling high-rate, cost-effective production suitable for industrial implementation. In reality, the boundary between nanotechnology and microtechnology may be blurred, but there is a degree of commonality in the techniques and equipment involved in both-but they are, in essence and in application, very different. It is at the nano not the micro scale that the physical and chemical properties of materials change. Micromachining is essentially a top-down technology but at the nanoscale, either top-down or bottom-up techniques can be used, and the latter are significantly different. Many products require a variety of top-down processes for their manufacture. For example, the common CD has data pits about 500 nm wide and 125 nm deep formed in a plastic disc. The read-write heads are very precise mechanisms that require a number of electro-mechanical processes.

Mass production is a major issue in MNT, as a large number of small parts are a common requirement of industry. Most of MNT have not yet been developed sufficiently to produce thousands of parts needed for microproducts. Modification of existing micro & nanomachining methods and new concepts for mass production based on them are major targets for future research and development. The present need for smallness of parts stems mainly from such requirements: greater compactness in the utilization of space and portability and reduce energy consumption. The mechanical and electrical devices that make up these items need to be produced in ever-decreasing sizes, with tightly specified dimensions and accuracies. Although these miniature devices may be manufactured by various procedures, their shaping by means of material removal constitutes a major means of production. Established and recently developed methods of machining continue to be investigated for the shaping of such parts to specified small dimensions. The term "micro and nanomachining" has thus emerged and is generally used to define the practice of material removal for the production of parts having dimensions that lie between 1 and 999 μm, although an upper limit of 500 μm has recently been considered to set the border between micro- and macromachining. This eBook is concerned with the technology of micro and nanomachining of materials utilized in engineering practice.

1.1. Single Point Diamond Turning (SPDT)

It is originated from 1950s that the ultraprecision machining of precision parts using natural diamond cutting tool came into being. At the beginning, the SPDT mainly used for machining simple shaped structure such as cylindrical surface, flat surface and spherical surface and the surface roughness (R_{max}) no larger than 0.1 μm. Later, the non-spherical surface reflector and the large scale reflector have also been gradually machined using MNT to acquire small form accuracy and surface roughness. The use of diamond cutting tools has increased in importance as tighter tolerances and greater surface integrities are required for high-value components. Ultraprecision cutting tools need to be hard and sharp and to have enhanced thermal properties in order to maintain their size and shape while cutting. Advantages offered by diamond include:

1. Crystalline structure, which enables very sharp cutting edges to be produced,

2. High thermal conductivity, the highest of any materials at room temperature,

3. Ability to retain high strength at high temperatures,

4. High elastic and shear modulus, which reduce deformation during machining.

Ultraprecision turning technology use diamond cutting tool (as shown in Fig. (**1-3**)) can be divided into two categories, namely, single piece large sized ultraprecision parts machining and mass small precision parts machining. The most active groups developing and using large sized parts ultraprecision turning technology are the Lawrence Livermore Laboratory at the University of California-Berkeley, in the United States who also leading the way in the world, Precision Engineering Center at the Cranfield University, in the UK, Precision Engineering Center at the Tohoku University, in the Japan. The highest level of the large-sized ultraprecision machine tool was developed in 1984 by LLNL in the USA, namely, the Large Optics Diamond Turning Machine (LODTM), which can manufacture non-spherical workpiece weighs 1,360 kilogram, 1,625 mm in diameter, 0.0125 μm in machining accuracy, 0.0045 μm in surface roughness (R_a).

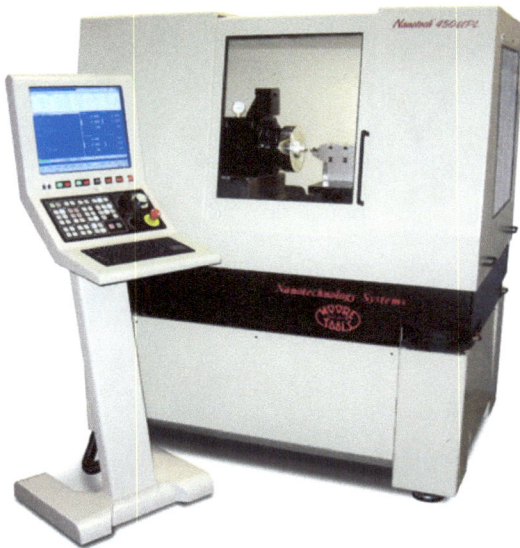

Precision Ground Silicon Carbide

Figure 1: Nanotech 450UPL and its product (The Nanotech 450UPL is a larger capacity ultra-precision machining system suitable for both single point diamond turning and deterministic micro-grinding of optical components).

Two off-axis parabolic mirrors were finish-turned on LODTM in an electroless nickel coating on aluminum substrates

Figure 2: Large optical diamond turning machine (LODTM-The LODTM is a precision, vertical-axis lathe that can machine and measure parts up to 1.5m in diameter, 0.5 m in length and 1350 kg mass).

Hard turning on DeltaTurn 40

Figure 3: Cranfield Precision DeltaTurn 40.

Although diamond is the hardest materials in the world, it may be chemically attacked by ferrous materials at high temperatures, and is generally unsuitable for the machining of steels and nickel alloys. This is because of the very high wear rate of the diamond which results in nonviable tool costs. More recently diamond machining has been used for the machining of nonferrous metals such as aluminum and copper, which are difficult materials on which to obtain a mirror surface by grinding, lapping, or polishing. This is because these metals are relatively soft and the abrasive processes scratch the finished surface and, furthermore, are unable to produce high levels of flatness at the edges of the machined surface. Presently, there are many small and medium-sized precision parts manufactured in mass production such as photosensitive drum, polygonal mirror, hard disk, spherical reflector mirror and so on. The workpiece materials being used are copper, aluminum and their alloy, electroless nickel plating layer, plastic and brittle materials, ferrites materials, ferrous metals and so on. Presently, the materials machined by SPDT has extended from traditional aluminum (copper) to difficult-to-cut materials and non-metal hard brittle materials, the research effort has extended from single cutting to develop systematic precision engineering with feed-back control and surface modification.

1.2. Ultraprecision Loose Abrasives Machining (ULAM)

Abrasive processes have been employed in manufacturing for more than 100 years although the earliest practice can be traced back to Neolithic times. Lapping and polishing (as shown in Fig. (**4**)), being the main technology of MNT, generally occur by the sliding frictions between particles and a surface. The lap or polisher travels across a work surface against which particles of sand or mud-type slurry are forced to the point of contact. Polishing involves only one or two of the abrasive mechanisms aforementioned. This widely used finishing process is one in which parts are finished on a plate covered with an abrasive pad. The polishing pad comes in a variety of thicknesses and hardness. Abrasive is often supplied in a paste suspension, but can be continuously fed suspended in a liquid carrier. Only two material mechanisms occur with this form of polishing-rolling and sliding. Abrasive is not embedded into the pad, therefore the microcutting mechanism is not active. Other types of mechanical polishing use different mechanisms for material removal. One type uses abrasive embed into the plate or a pad, but no additional abrasive is applied to the polishing surface. With this type, material removal is only through the microcutting abrasive mechanism. For all types of polishing, generally two abrasive mechanisms are involved.

Lapping on the other hand, incorporates all three abrasive mechanisms: rolling abrasive, sliding abrasive, and microcutting abrasive. The plate is not covered with a pad and therefore contributes in the material removal process. With typical lapping operations, abrasive is forced into the lap plate, called charging, and the parts are lapped with continuously supplied abrasive suspended in a liquid medium.

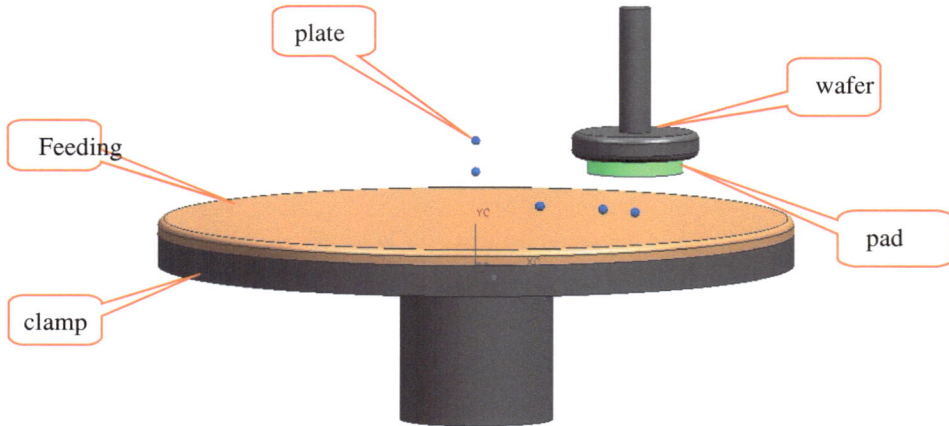

Figure 4: Illumination the working conditions of CMP.

With both processes, material removal is by rolling abrasive, sliding abrasive, or microcutting embedded abrasive. The action of sliding abrasive and rolling abrasive are implied. They are mechanically similar in their cutting action except that sliding abrasives are more plate-like and behave like tiny scrapers. Microcutting abrasives are abrasives that have embedded into the lapping plate and act like small cutting tools.

1.3. Laser Materials Micromachining (LMM)

Laser light can be focused on a precise area with very high heating because of its high intensity of electromagnetic energy flux and high spatial coherence. Therefore, materials processing has becoming one of the major applications of laser [2-3]. Comparing with conventional machining process, laser materials micromachining has the advantages of non-contact, good machining quality, high flexibility and easiness of automation. Laser cutting, laser hardening and laser remelting are thermal related process while laser alloying, laser cladding and laser dispersing are thermo chemical process. Laser materials micromachining is based on the interaction of laser light with materials and the reaction products escape as gas or small particles. As a result of a complex process, small amounts of material can be removed from the surface of the solid. Generally, two different phenomena may be identified: pyrolithic (thermal) and photolithic processes. In both cases short to ultrashort laser pulses are applied in order to remove small amounts of material in a controlled way. Pyrolithic processes are based on a rapid thermal cycle, heating, melting, and (partly) evaporation of the heated volume. In the case of photolithic processes the photon energy is sufficient for direct breaking of the chemical bonds in a wide variety of materials. It is applied mostly on polymers by use of ultraviolet lasers in wavelengths of 157 to 351 nm.

Laser micromachining belongs to materials removal technique which consists of chemical or electro-chemical process, thermal process and mechanical process. Laser micromachining includes a wide range of processes where material is removed accurately but the term is also used to describe processes such as microjoining and microadjustment by laser beam. Most applications are found in the electronics industry in high-volume production. The earliest industrial applications occurred in the 1960s in the cutting of trim grooves on conventional resistors and drilling small holes in diamonds. In the 1970s laser spot welding was applied to the production of lamps and parts of television monitors. The 1980s saw the beginning of laser micromilling and laser ablation with excimer lasers, while in the 1990s laser microadjustment was developed for use in industry. With the development of new lasers such as ultrashort pulsed lasers and passively Q-switched microlasers, new applications continue to arise. From the beginning of laser technology in the 20th century a reduction in size by a factor of two every seven years has been observed. The cost of production equipment is growing much faster: the complicated optics of a step-and-repeat camera for semiconductor production is now over a million dollars. Nevertheless the cost of products is being reduced, owing to the higher production volume.

In laser machining, the laser light works as a high energy heating source by thermal ablation mechanism, photochemical ablation mechanism or both to melt and/or vaporize the volume of the material. It is a non-contact processing. Combined with modern numerical control and CAD technology, the machining rate and precision increase significantly. It is used in variety industry fields, particularly in the processing of difficult-to-machine materials such as hardened metals, ceramics and composites. The mechanism of laser beam interaction and material removal is shown in Fig. (**5**). Laser energy is focused on the material surface and partly absorbed. The absorptivity depends on the material, the surface structure, the power density, and the wavelength. With a CO_2 laser about 20% is absorbed with laser micromachining while with shorter wavelengths (Nd:YAG and excimer lasers) 40% to 80% is absorbed. The remaining part is reflected. Absorption occurs in a very thin surface layer, where the optical energy is converted into heat. The optical penetration depth is defined as the depth for which the power density is reduced to $1/e$ of the initial density. For steel this depth is on the order of 15 nm for CO_2 radiation or 5 nm for Nd:YAG radiation. The absorbed energy diffuses into the bulk material by conduction. For short pulses the heat flow is approximately one-dimensional. However, the time to melt is reduced by a factor of 100 to only 3 ns if the power density is increased tenfold. The high vaporization rate (vapor speeds have been reported in the range of 3 to 10 km/s) causes a shock wave and a high vapor pressure at the liquid surface considerably increases the boiling temperature. Finally the material is removed as a vapor by the expulsion of melt, as result of the high pressure and by an explosive like boiling of the superheated liquid after the end of the laser pulse. In metals a rim of resolidified material caused by laser micromachining is clearly evident. In plastics, however, the process is quite different; here the material is removed by breaking the chemical bonds of the macromolecules, and is dispersed as gas or small particles and no melt is found.

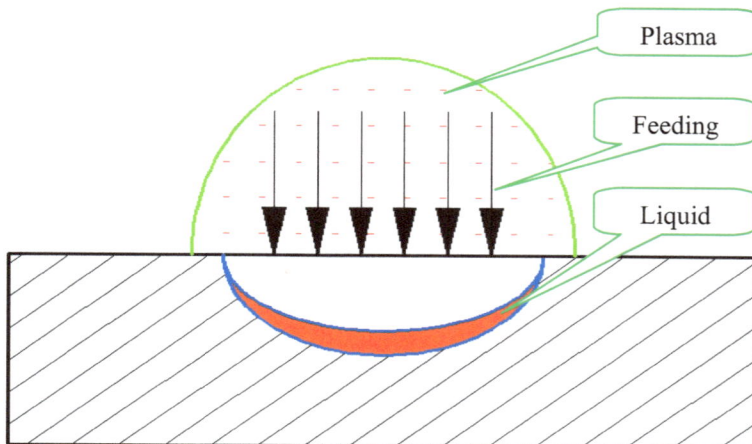

Figure 5: Schematic of working principles of laser beam.

Electromagnetic waves interact with the particle on the surface of the material, the electron will re-radiate, or be constrained by the lattice; if enough energy is put into the material the lattice breaks down and the material begins to melt. Further heating causes evaporation and plasma formation to occur. When laser radiation hits a surface it is absorbed, transmitted or reflected depending on the material. The laser radiation that interacts with the particle has a magnetic and an electric component. When the radiation passes over a small elastically bound the charged particle is set in motion by the electric field. This induced force is so small that it cannot affect the nucleus but can affect the electrons. If the electrons were left to vibrate there would be no net gain in its energy, *e.g.*, its motion would be in the form of a sinusoidal wave, a few positive and a few negative motions resulting in zero energy change. However, if the electrons are involved in a collision, the path will be upset and they will gain some energy. If the radiation is at a lower potential than the ionization energy for the particles then no absorption occurs.

Currently used industrial lasers are mainly CO_2, Nd:YAG, excimer, argon ion, and copper vapor types. General-purpose machining equipment consists of a stationary laser beam with a product holder on a horizontal xy-stage and a lens capable of moving in the vertical direction. The solid state Nd:YAG laser is the main vehicle for micromachining applications. The energy is pumped by flash lamps into the Nd:YAG rod. The laser beam with about 6 mm diameter can be focused by lenses directly on the surface to spots of diameter 50 μm for fine drilling or cutting, to about 0.5 mm for spot welding.

The laser-material interaction consists of a set of physical steps each characterized by its typical time constant. The laser energy is first transferred to the electrons, especially in the case of metals. The electrons will transfer the energy to the lattice and finally, within the lattice the heat is distributed further by atomic lattice collisions. The first step, the absorption of a photon by an electron, requires about 10^{-15} s (1 femtosecond). The relaxation time of a high-energy electron, that is, the time to transfer the energy to the lattice, is about 10^{-12} s (1 picosecond). The time to diffuse the heat in the lattice by thermal conduction, over a distance equal to the optical penetration depth, is also on the order of 1 picosecond. Three different processes may be identified, based on these characteristic times.

1.3.1. Femtosecond Ablation

In this case there is no energy transfer to the lattice during the pulse; all energy is stored in a thin surface layer. If this energy is more than the specific heat of evaporation there will be vigorous evaporation after the pulse. The energy is transferred from electrons to the lattice (after the laser pulse) in a picosecond. This transfer converts the layer in a dense vapor or plasma, which expands rapidly. No time is available for heat transfer to the lattice during this series of processes. The outcome is a very precise and pure laser ablation of metals; this result has been demonstrated experimentally but has not yet been well established in production.

1.3.2. Picosecond Ablation

With picosecond pulses the pulse length is on the same order as the time to transfer the energy from electrons to the lattice. The lattice temperature at the end of the pulse is approximately equal to the femtosecond ablation. Although the heat conduction into the lattice may be neglected there will be a considerable heat flow by the free electrons during the pulse. This results in the formation of a melted zone inside the material. At the surface there is a direct solid-vapor or solid-plasma transition but deeper in the material a liquid phase is present. This condition reduces the precision of the ablation of metals compared to femtosecond pulses.

1.3.3. Nanosecond Ablation

In terms of the thermal processes during laser-material interaction the nanosecond pulses have to be considered as long pulses. The absorbed laser energy first heats the work specimen to its melting point and then to the vaporization temperature. During the interaction the main energy loss is by heat conduction into the solid. The threshold influence for long pulses can be estimated in the same way as for ultrashort pulses. Generally droplets or crater walls are observed around the machined area in the nanosecond domain.

Laser machining is a thermal or photochemical process which makes it appropriate for hard & brittle materials and composites. There are no mechanically induced material damage, tool wear and machine vibration.

Furthermore, the material removal rate for laser machining is not limited by constraints such as maximum tool force, build-up edge or tool chatter. Laser machining can eliminates the workpiece transportation that is necessary for processing parts with a specialized machine. So it can be used for drilling, cutting, welding and heat-treating process on a single machine. On the other hand, there are some disadvantages about laser machining. In most laser machining techniques, the removal of materials occurs by melting or vaporizing which makes the laser machining a significantly high energy inputs and processing times technique than mechanical technique. During laser machining metals, the temperature of the point where the laser focus is very high and the heat conduction will create a heat affected zone in the vicinity of the erosion front and change the surface integrity.

Laser nanofabrication is at a very early stage of development at the moment and will probably need to be combined with other processes to form hybrid nanofabrication and nanomanufacturing processes. A large number of researchers are currently investigating the development of hybrid forms of laser nanofabrication. The use of lasers to fabricate products at the microscale is well established. CFD studies of supersonic jets interacting with holes of varying aspect ratio have highlighted complex shock wave structures present during laser micro-machining that enhances the shear stress distribution in the molten pool that is required to physically remove molten material and send it into the jet stream above the surface of the material. Experimental observation has revealed reduced etch rates at high gas pressures, which is coincident with dense plasma formation within the shocked gas stream above the surface of the workpiece. A reduction of gas pressure to near atmospheric pressure reduced detrimental plasma effects and reduced the height of re-cast layers once dominant at high pressure. A simple design rule used for laser micro-machining is to reduce the length-to-diameter ratio of the nozzle, and use the nozzle with shielding gas pressures in the range 0 to 0.5 bar in order to prevent the formation of shock waves in the dynamic plasma field that is responsible for the formation of re-cast layers. The development of processing at the microscale continues to be dominated by the application of femtosecond pulsed processing of engineering materials. However, the development of attosecond pulsed lasers may eliminate the problems created by the formation of a plasma that accompanies the more traditional lasers currently used for microfabrication. Nanofabrication processes that use lasers to create useful nanofeatures are still in their infancy of development. Although the rapid strides made in laser manipulation at the nanoscale may see the development of laser-based nanomanufacturing processes in the not too distant future.

1.4. Ion Beam Materials Micromachining

Ion Beam (IB) is a precise and accurate instrument which can be applied to the fabrication of microelectronic and optoelectronic devices such as direct-write lithography, circuit modification, failure analysis of Integrated Circuits (ICs), Transmission Electron Microscopy (TEM) sample preparation, maskless implantation, ion beam assisted etching and micromachining. For those practical applications of IB micromachining like TEM sample preparation and failure analysis of ICs, the removal of sample materials is mainly done by physical sputtering of ion bombardment. The ion beam (usually Ga+ ions or other liquid metal ion sources) scans through the desired location and mills a very small area around the region of interest, without causing damages to the other areas of the sample. Because IB technique is a maskless and resistless process, some researchers have demonstrated the capability to process precise micromachined laser facets at the areas of semiconductor laser devices. Ion beam machining (Fig. (**6**)) takes place in a vacuum chamber, with charged atoms (ions) fired from an ion source towards a target (the workpiece) by means of an accelerating voltage. Ion beam machining (IBM) is associated with the "sputtering" phenomenon first reported by Grove in 1852 [4]. While investigating the electrical conductivity of gasses, Grove discovered that metallic substances had become deposited on the glass walls of the glow discharge tube that he was using. He inferred that metal atoms had been removed from the surfaces of the electrode, and subsequently had adhered to the walls of the glass tube. Later the mechanism underlying Grove's finding was established as the ejection of atoms from a surface when it is bombarded by other ions.

An ion beam machine has several main components:

1.   A plasma source that generates the ions;

2.   Extraction grids for removing the ions from the plasma and accelerating them towards the substrate (or specimen);

3.   A table for holding the specimen.

Orsay Physics Canion 31 Plus UHV FIB on a TOF-
SIMS 6600 from Physical Electronics

Figure 6: Ion beam workstation.

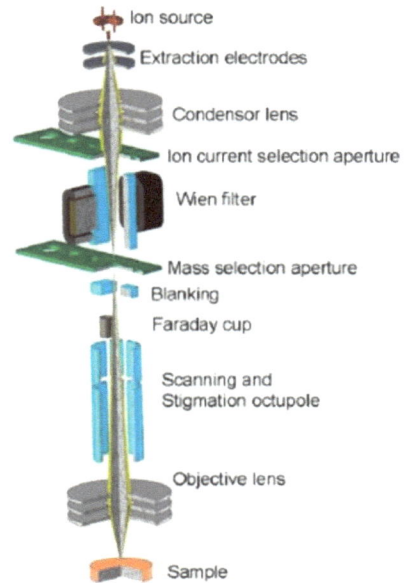

Basically, there are some key parts in the ion beam micromachining system as follows:

1.4.1. Ion Source

For the removal of an atom from the surface by impingement of an ion, or ions, a source of ions is required that should produce a sufficiently intense beam with an acceptable spread in its energy.

1.4.2. Plasma Source

A heated filament, usually tungsten, acts as the cathode, from which electrons are accelerated by means of a high voltage (above 1 kV) towards the anode. During the passage of the electrons from the cathode to the anode, they interact with argon atoms in the plasma source (which is sustained by keeping the gas pressure at about 10^{-4} torr). A magnetic field, obtained from an electromagnetic coil or a permanent magnet, is often applied between the anode and cathode to make the electrons spiral. Spiralling increases the path length of the electrons and hence increases ionization.

1.4.3. Extraction Grids

The ions are removed from the plasma by means of extraction grids. The grids are normally made of two or three arrays of perforated sheets of carbon or molybdenum; these materials can withstand erosion by ion bombardment. The perforations in each of the sheets are aligned above one another. The shape of the holes and the spacing of the grids are significant elements when ion source systems are being designed to give the best conditions of ion current and grid erosion.

1.4.4. Substrate Mounting

When the ions have been removed from the source, they "drift" in a field-free region to the component, specimen, or substrate which is to be machined or milled. The specimen is usually mounted on a water-cooled table that can be tilted through an angle of $0°$ to 90. The specimen is separate from the plasma. Machining variables such as acceleration, flux, and angle of incidence can all be independently controlled.

The interactions of the ions with atoms are now considered. The size of an ion is normally comparable to that of an atom. When an ion strikes the surface of a material its collision with an atom there often occurs in a direction that is normal to the surface. If the mass of the ion is less than that of the atom of the surface, the former will bounce back, away from the surface, and the atom will be driven in a direction farther into the material. For higher ion energies a yield of approximately 0.1 to 10 atoms per incident ion is representative of IBM. The effects influence a wide range of process conditions, not only incident ion energy but also the angle of incidence, ionic and atomic periodicity, and heat of sublimation. The rate of material removal (or etching) can be enhanced at grain boundaries and can be affected by crystallographic orientation.

Presently, the structure dimension as small as 100 nm can be realized by IBM, and that features of size less than 10 μm are obtainable. The slope of the walls of the machined surface and its surface finish are determined by the angle of incidence of the ion beam which is fully controllable. They claim that the process variables can be monitored to an accuracy of $\pm 1.0\%$, with a repeatability of $\pm 1.0\%$. Surfaces can be textured by IBM, which can produce cone- and ridge like configurations, on the order of 1 μm in size. Smoothing of a surface to a finish of less than 1 μm can also be achieved by IBM. A smooth surface can undergo ion beam machining without significant increase in its roughness.

Today, much work has been published on the use of microfocused ion beams for machining, although this technique is almost entirely confined to academic research. Currently, the IBM technology can be used for smoothing of laser mirrors and for modifying the thickness of thin films and membranes without affecting the surface finish such as Ion Beam Texturing, Ion Beam Cleaning, Shaping, Polishing, and Thinning by IBM, Ion Milling and so on. For production applications laser ablation is usually preferred for small spots, and plasma etching for large areas. Much of future ion beam technology is likely to deal increasingly with deposition. In this method, the material to be deposited is ionized inside the source and then directed in the form of an ion beam to the surface on which it will condense. The technique is used to study growth of thin films for deposition of metals such as pure iron and semiconductor layers of silicon. An ion beam sputter deposition technique is also reported, the material being sputtered from a target by an energetic ion beam and collected onto the substrate. Films of oxides and nitrites can be deposited by reactive ion beam sputtering. New developments on radio frequency or microwave ion guns continue. Most trends are concentrated on automatic control of the source for utilization in complex processes, neutralization of the ion beam in reactive gas environments, increases in ion density, and very low levels of contamination.

1.5. Electron Beam Materials Micromachining

Electron beam machining (EBM) is a process where high-velocity electrons concentrated into a narrow beam are directed toward the work piece, creating heat and vaporizing the material. EBM can be used for very accurate cutting or boring of a wide variety of metals. Surface finish is better and kerf width is narrower than those for other thermal cutting processes. The main components of an Electron Beam Machine (EBM) are shown in Fig. (**7**). They are housed in a vacuum chamber, evacuated to about 10^{-4} torr. The source of electrons is an "electron gun", which is basically a triode consisting of a cathode, a grid cup negatively biased with respect to the cathode, and an anode at ground potential. The cathode is usually made of a tungsten filament, which is heated to between 2500℃ and 3000℃ in order to emit the electrons. A measure of this effect is the emission current, the magnitude of which varies between 25 mA and 100 mA. Corresponding current densities lie between 5 Acm^{-2} and 15 Acm^{-2}. This quantity, however, is determined by a range of factors, including the type of cathode material and its temperature (see below). The size of the emission current is also influenced by a high voltage, usually about 150 kV, which is applied between the cathode and anode in order to accelerate the electrons in the direction of the workpiece.

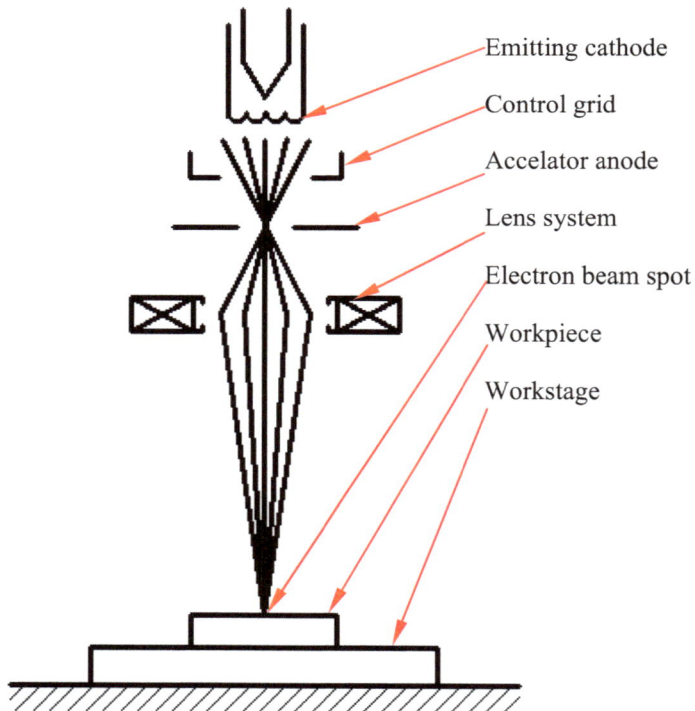

Figure 7: Schematic of electron beam machining.

After acceleration, the electrons are focused by the field formed by the grid cup so that they travel through an aperture in the anode. On its exit from the anode cavity, the electron beam is refocused by a magnetic or electrostatic lens system. By this means, the beam has its direction towards the workpiece kept under control. The electrons maintain the velocity imparted by the accelerating voltage, until they strike the workpiece specimen, over a well-defined area, typically 0.025 mm in diameter. There the kinetic energy of the electrons is rapidly translated into heat, causing a correspondingly rapid increase in the temperature of the workpiece, to well above its boiling point. Material removal by evaporation then occurs. With power densities on the order of 1.55 $MWmm^{-2}$ involved in EBM, virtually all engineering materials can be machined by this technique. Fig. (7) shows details of an industrial electron beam machine. This particular unit can be used for welding as well as machining. Accurate manipulation of the workpiece coupled with precise control of the beam can yield a process that can be fully automated.

A useful account of the theory underlying EBM has been presented by McGeough [5] on which the following discussion is based. This section provides a useful background to the work on electron beam technology for high-resolution lithography. The emission current is a significant variable in EBM. When the voltage gradient in front of the emitter is sufficiently high to draw off the electrons, a condition known as "temperature limited emission" occurs, giving rise to a maximum emission current density. When the voltage gradient is less than that above, the emission of electrons is hindered by a negative space charge in the vicinity of the cathode. Mutual electron repulsion takes place there. The emission current is then known as space-charge limited. In the region where the beam of electrons meets the workpiece, the energy is converted into heat. The way in which the focused beam penetrates the workpiece is still not completely understood, owing to the complexity of the mechanisms involved; however, it is known that the workpiece surface is melted by a combination of electron pressure and surface tension. The melted liquid is rapidly ejected and vaporized to achieve material removal. The temperature of the workpiece specimen outside the region being machined is reduced by pulsing the electron beam and the pulse frequencies seldom exceed 104 Hz.

The energy associated with a single pulse may be calculated from the product of the accelerating voltage, the emission current, and length of pulse. The corresponding power density has a significant effect on the material removal rate, through the rate of heating of the workpiece. An early attraction of EBM was the comparatively large depth-to-width ratio of material penetrated by the beam with applications of very fine hole drilling. The depth to which the electron beam penetrated the material received close attention. The analytic expressions produced have often been complicated.

Electron beam machining rates are usually evaluated in terms of the number of pulses required to evaporate a particular amount of material. Two types of pulse numbers are used. The volume number is used for evaluating slotting by EBM, that is, "cutting-off" or "cutting-into" materials, and is given by the ratio of the mass to be removed to that due to a single pulse. A linear number is adopted for hole sinking, and is the ratio of the depth of the hole required to that sunk by a single pulse. Studies of the EBM of different metals have revealed that their boiling point and thermal conductivity play a significant role in determining how readily they can be machined. Properties such as electrical conductivity of the material are additional factors. The complexity of the relationship between the material properties and machinability renders difficult any fully quantitative analysis of removal rates in EBM. The quality or surface roughness of the edges produced in EBM depends greatly on the type of material. Local pitting of the surface is a common occurrence, the extent of which is influenced by the thermal properties of the workpiece, and by the pulse energy or charge. Surface roughness can increase with pulse charge for a range of common materials such as nickel, titanium, carbon, gold, and tungsten. Applications lie in the following main areas: (a) drilling, (b) perforating of sheet, (c) pattern generation associated with integrated circuit fabrication (with which milling is also associated), and (d) texturing.

Steigerwald and Meyer [6] gave early consideration to EBM for hole-drilling. They concluded that improved reproducibility, greater working speeds, and deeper holes of accurately controlled shapes were needed. Later Boehme [7] discussed drilling applications with electron beam machines fitted with systems for numerically controlling the beam power, focus, and pulse duration. As a result cylindrical and other configurations, such as conical- and barrel-shaped holes, of various diameters were drilled with consistent accuracy at rates of several thousand holes per second. Drilling of inclined holes, at an angle of 15°, was also investigated.

The sheet to be perforated is usually lined with an auxiliary material. The electron beam first penetrates the sheet forming a vapor channel within the fused material, and then enters the auxiliary lining. An eruption of vapor occurs, causing ejection of molten material. For perforation by EBM to be economically acceptable, 104 to 105 holes per second have to be produced. Thus single pulses lasting only a few μs are needed. In some applications the sheet or foil metal is stretched on a rotating drum, which is simultaneously shifted in the direction of its axis. Rows of perforations following a helical line are thereby produced. Manipulators capable of linear and rotary movement in as many as four axes are used, especially for perforation by EBM of jet engine components.

Electron beam machining belongs to non-contact machining which result in little macroscopic deformation and tool wear. Electron beam can feed at any rate under the electrical field or magnetic field which makes it easy to be controlled. There is no pollution in electron beam machining technique which makes it fit for machining pure semiconductor material or easy oxidated metals. The special feature of electron beam makes it an ideal machining tool which can be used for drilling, cutting, welding and photolithography.

1.6. Ultrasonic Micromachining (USM)

Ultrasonic vibrations are used in many fields of industry for applications such as cleaning, welding, and machining (Fig. (**8**)), including assistance to cutting or grinding, nondestructive control, mixing, and separation. Applications involving ultrasonic waves can be placed in two groups:

The transmitted power is small (mW to W); it is mainly used for nondestructive applications, in which ultrasonic waves must not damage either the item or a live body part that is to be treated. The transmitted power is an important factor (W to kW), and the waves are used to cause modifications to the part. Particles affect both part and sonotrode: material removal takes place at the workpiece; wear occurs on

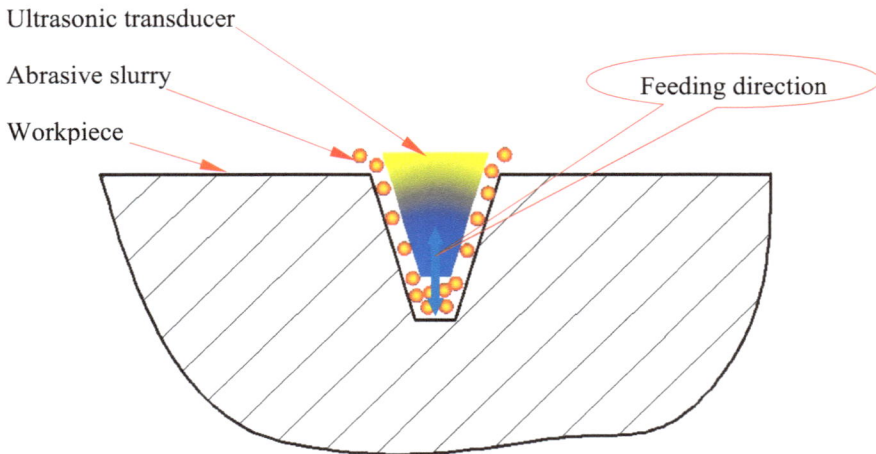

Figure 8: Principle of ultrasonic machining.

both the sonotrode and particles. The process is therefore characterized by removal rate on the workpiece, sonotrode wear, and abrasive wear. Ultrasonic machining can deal efficiently with brittle materials. The sonotrode should be made from materials that can resist wear: they should be either ductile (aluminum alloys, steel, titanium alloys, nickel alloys) or extremely hard (diamond). The abrasive particles have to be harder than the workpiece material: aluminum oxide, silicon carbide, boron carbide, and diamond are used. The USM process is able to machine any material, but is more efficient on brittle materials. Materials respond differently to the process.

There are many process variables which have a great impact upon materials removal in the USM technology. The effect of abrasive particles also depends on their shape and dimension. It is difficult to adopt shape as a parameter, and main diameter is often taken as the particle dimension parameter. The influence of sharpness of edges on grains is not discussed here. In general, it has been found that increasing the main diameter d of the impinging particles leads to a rise in the amount of eroded material. Sometimes there is a major effect: material removal rate can be proportional to d^4. Often in ultrasonic machining, a behavior is observed that is composed of first a very significant increase for small diameters, then no further increase, and eventually a decrease. This behavior can be explained by the close relationship between this parameter and amplitude of vibrations, and for conditions where amplitude is either not sufficiently high to project grains with adequate efficiency, or too great.

Amplitude and frequency of vibrations are significant variables. Frequency greatly influences the efficiency of the process. Removal rate increases with frequency. But it is noted that increasing frequency leads to higher demands on the acoustic system, giving rise to greater heating and thus smaller fatigue life of components; increasing frequency requires smaller amplitude, which does not benefit removal rate. At 20 kHz, an amplitude of 20 to 50 μm can be obtained; at 40 kHz, it is limited to 5 to 20 μm for the same mechanical resistance. The machine conditions have to be carefully selected in order for the working gap to provide for effective action by the abrasive particles. These conditions in turn depend on the other variables and on the material to be machined. In sinking, the gap between sonotrode and part is adjusted by static force; in contouring, there is no adjustment of gap, which depends on the choice of the variables used. In sinking, there is an optimum value of static force. A smaller static force gives rise to a larger gap, particles are not adequately efficient, a higher static force is associated with a smaller gap, and particles have a restricted motion. In contouring, the main variables are depth of cut and lateral feed speed. An increase in these two variables causes more material to be removed per unit of time and thus decreases the working gap. In this case also, an optimum has to be found for selecting all variables that are interdependent.

The most significant condition is grain diameter which has a major effect on quality. The smaller the grain, the lighter is the impact, and the better the quality. A relationship can be found between grain diameter and crater dimensions: crater diameter is about $d/3$; crater depth is approximately $d/10$. For the achievement of higher quality, very fine grains should be used, and of course values of other variables have to be appropriately adapted: smaller amplitude, smaller static force (sinking), smaller depth of cut, and larger lateral feed speed (contouring).

In general, the design accuracy entails both roughing and finishing, since quality can seldom be obtained in a single operation. Roughing is performed with large grains (20 to 120 μm) to give sufficient removal rate; finishing is achieved through grains fine enough (0.2 to 10 μm) to obtain the desired quality. Drilling of very small holes is performed in a single operation. Tool wear is of major consideration for accuracy, since it affects both tool geometry and dimension. The use of metal alloys leads to noticeable wear, for example, from less than 1% up to 50% when machining, respectively, graphite and glass, and silicon nitride. Carbide provides greater resistance, but still gives some wear. The solution to tool wear rests with using diamond, which has been proved to give a wear so small that it could be considered nonexistent. Micro-drilling with diamond as abrasive gives a 1% maximum wear ratio, whereas carbide and steel can lead to a 75% ratio when machining hard ceramics.

The main practical relevance of using ultrasonic vibrations is to obtain a decrease in machining forces. This condition is obtained mainly by the reduction in friction coefficient observed when one of the two surfaces in contact is vibrated. Other advantages are observed: ultrasonic vibrations facilitate evacuation of chips and provide cleaning of the tool. This procedure is used in either grinding or cutting, for producing parts from metal alloys or ceramics, and also for medical applications (dentistry, surgery).

1.6.1. Rotary Ultrasonic Machining

This technique involves the vibration of a small grinding tool, and is used for machining brittle materials (glass, ceramics, stone). A large reduction in machining forces is obtained (two to four times), which permits either an increase in machining speed or a decrease in machining force, a useful condition for machining with very fragile tools. Removal rate can be two to four times higher depending on the material of the component. It is possible to drill 1 mm diameter holes to a 75 mm depth in glass and to machine narrow slots in ceramics. Ultrasonic vibrations give a reduction in drill wear, by evacuating chips blocked between abrasive grains, and also worn grains. Productivity is drastically increased by ultrasonic assistance.

1.6.2. Ultrasonic Assistance Cutting

In cutting, the main required effect of ultrasonic assistance is the reduction in friction among tool, chips, and part. The first effect is a large reduction in cutting forces, in turning as well as in drilling, which has been observed for metal and ceramic materials. The second benefit is a significant improvement in surface quality, the machined surface showing less tearing when ultrasound conditions are applied. This effect is particularly relevant in the treatment of materials that are difficult to machine which have an inherent tendency to adhere to the tool edge (the so-called "built-up edge" phenomenon): aluminum copper, stainless steel, nickel and titanium alloys.

As noted earlier, in most applications ultrasonic machining has been mainly used for producing small details on parts. Its physical principles make the process useful for machining brittle and hard materials, and disadvantageous for ductile materials. It has thus been used for graphite, glass, quartz, ceramics, precious stones, semiconductors, and diamond. The need to work at very high frequencies (at least 20 kHz) restricts the dimensions of the working end of the tool to 25 to 50 mm in general. This restriction means that ultrasonic machining can be applied to micromachining. To this end, since the result to be obtained on the part depends on tool dimensions and working variables, it is necessary to be able to use tools of size from as small as a few micrometers to one millimeter, and adjust the process variables: grains from 0.2 μm to 20 μm in diameter; amplitudes from 0.1 μm to 20 μm; and forces from 0.1 mN to 1 N.

1.7. Micromachining Using High-Resolution Lithography

During the last decade there has been a rapid development in microfabrication technology driven by the market need for low-cost consumer products such as portable telecommunications equipment, computers, and healthcare diagnostics. Much of the technology used for these is based on production of silicon semiconductors and microchips. Interest in non-silicon-based technologies started to grow back in the early 1980s with the development of a German fabrication process known as LIGA, an acronym for Lithography (Lithographie), electroplating (Galvanoformung), molding (Abformung). It originated at the Karlsruhe Nuclear Research Laboratory in Germany. Since then a number of groups, mainly in Germany and the United States have been active in developing the process to make precision microcomponents for a range of innovative products such as microspectrometers, fiber-optic wave guides, micro-reactors and microfluidic devices. A few of these have been manufactured on a large scale and placed on the market. LIGA was often used to fabricate the components, which were then integrated with others into the end product. High costs are often associated with the integration and packaging of the final product.

Lithography (Fig. (**9**)) is the process by which fine features are defined on a substrate. For conventional lithography the substrate needs to have a flat surface, and is typically a 200 mm diameter silicon wafer with a thickness of about 1.0 mm. Optical lithography is the most common form because it combines the accuracy required for current silicon circuits with considerable throughput capability. Electron beam lithography is capable of much higher resolution pattern definition but is a serial process and therefore much slower than optical lithography, which is a parallel process. X-ray lithography combines the higher resolution of electron beam lithography with the high speed parallelism of optical lithography. However, it is not as technologically developed as optical lithography and is currently a research, rather than a production, tool. The lithography process leaves a substrate coated with a finely patterned polymer film. Subsequent processing steps are needed to transfer this pattern to underlying materials that are usually metals, semiconductors, or dielectrics.

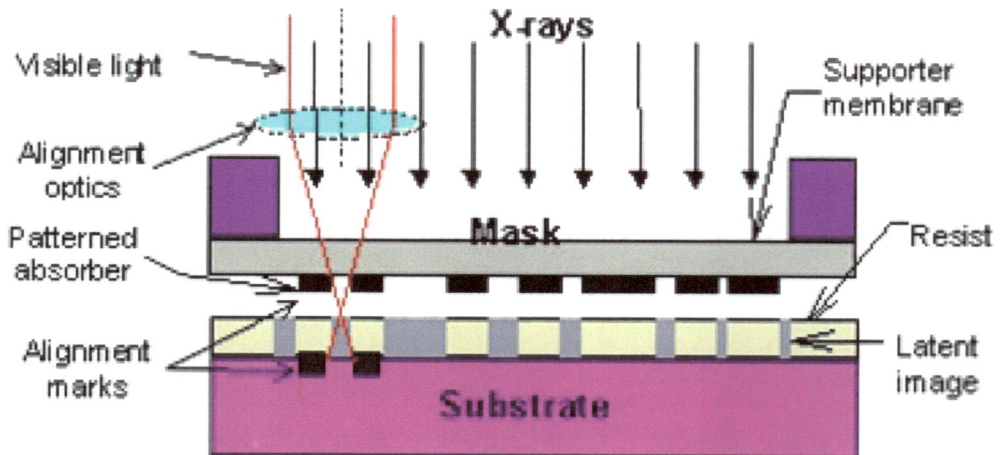

Figure 9: Schematic of projection optical lithography.

LIGA is a three-step process. Although here, we consider X-ray based lithography, UV-LIGA and Laser-LIGA techniques have also reached an advanced stage of development. UV-LIGA in particular has encouraged the production of a negative resist known as SU-8 that can also be used in X-ray lithography owing to its increased radiation sensitivity over more commonly used polymethylmethacrylate (PMMA) resists, thus reducing exposure time and subsequent costs. The penetrating power of X-rays compared to other longer wavelength radiations allows the fabrication of structures that have vertical dimensions from hundreds of microns to millimeters and horizontal dimensions as small as microns. These 3-D microstructures with high aspect ratios offer a range of microcomponents for many useful applications.

The first step using X-ray lithography involves exposing a thick layer of resist through a patterned mask to a high-energy beam of X-rays from a synchrotron. The pattern is etched into the resist substrate by the use of X-rays. A chemical solvent is used to dissolve away the damaged material, resulting in a negative relief replica of the mask pattern. Certain metals can be electrodeposited into the resist mold. After removal of the resist, a freestanding metal structure is produced. The metal structure may be a final product, or serve as a mold insert for precision plastic molding. Molded plastic parts may then be final products or lost molds. The plastic mold retains the same shape, size, and form as the original resist structure but is produced quickly.

1.8. Reactive Ion Etching (RIE)

Reactive-ion etching (RIE) is an etching technology used in micromachining. It uses chemically reactive plasma to remove material deposited on wafers. The plasma is generated under low pressure (vacuum) by an electromagnetic field. High-energy ions from the plasma attack the wafer surface and react with it whose principle is shown in Fig. (**10**).

A typical (parallel plate) RIE system consists of a cylindrical vacuum chamber, with a wafer platter situated in the bottom portion of the chamber as shown in Fig. (**10**). The wafer platter is electrically isolated from the rest of the chamber, which is usually grounded. Gas enters through small inlets in the top of the chamber, and exits to the vacuum pump system through the bottom. The types and amount of gas used vary depending upon the etch process; for instance, sulfur hexafluoride is commonly used for etching silicon. Gas pressure is typically maintained in a range between a few millitorr and a few hundred millitorr by adjusting gas flow rates and/or adjusting an exhaust orifice.

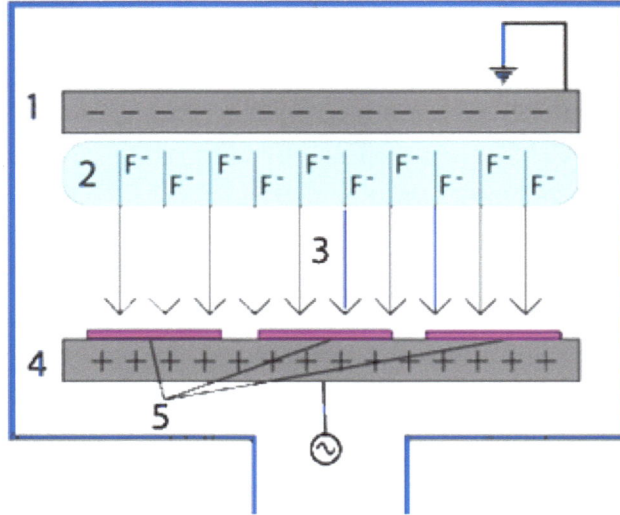

Figure 10: Principle of Reactive-ion etching, 1,4 - Electrodes; 2 - Accelerated ions; 3 - Electric field; 5 - Sample.

Other types of RIE systems exist, including inductively coupled plasma (ICP) RIE. In this type of system, the plasma is generated with an RF powered magnetic field. Very high plasma densities can be achieved, though etch profiles tend to be more isotropic. A combination of parallel plate and inductively coupled plasma RIE is possible. In this system, the ICP is employed as a high density source of ions which increases the etch rate, whereas a separate RF bias is applied to the substrate (silicon wafer) to create directional electric fields near the substrate to achieve more anisotropic etch profiles. Etch conditions in an RIE system depend strongly on the many process parameters, such as pressure, gas flows, and RF power. A modified version of RIE is deep reactive-ion etching, used to excavate deep features.

1.9. Atomic Layer Deposition

Atomic layer deposition (ALD) is a vapor phase thin film deposition technique based on sequential self-limiting surface reactions. In a typical ALD process, two reactants are introduced alternatively to a substrate, resulting in formation of a single atomic layer each exposure cycle. Repeating the surface-limiting reaction cycle allows ultra-uniform nanoscale films to be formed with precise thickness control over complex 3D surfaces. Similar self-limiting surface reactions have been used for the deposition of polymer or organic-inorganic hybrid films. ALD is a self-limiting (the amount of film material deposited in each reaction cycle is constant), sequential surface chemistry that deposits conformal thin-films of materials onto substrates of varying compositions. Due to the characteristics of self-limiting and surface reactions, ALD film growth makes atomic scale deposition control possible. ALD is similar in chemistry to chemical vapor deposition (CVD), except that the ALD reaction breaks the CVD reaction into two half-reactions, keeping the precursor materials separate during the reaction. By keeping the precursors separate throughout the coating process, atomic layer control of film growth is can be obtained as fine as ~0.1 Å (10 pm) per cycle. Separation of the precursors is accomplished by pulsing a purge gas (typically nitrogen or argon) after each precursor pulse to remove excess precursor from the process chamber and prevent "parasitic" CVD deposition on the substrate.

1.10. Self-Assembly

Self-assembly is a type of process in which a disordered system of pre-existing components forms an organized structure or pattern as a consequence of specific, local interactions among the components themselves, without external direction. When the constitutive components are molecules, the process is termed molecular self-assembly. Self-assembly process can be classified into two types, static or dynamic process. In static self-assembly, the ordered state forms as a system approaches equilibrium, reducing its free energy. However in dynamic self-assembly, patterns of pre-existing components organized by specific local interactions are not commonly described as "self-assembled" by scientists in the associated disciplines. These structures are better described as "self-organized". Self-assembly processes can be observed in systems of macroscopic building blocks. These building blocks can be externally propelled or self-propelled. Since the 1950s, scientists have built self-assembly systems exhibiting centimeter-sized components ranging from passive mechanical parts to mobile robots. For systems at this scale, the component design can be precisely controlled. For some systems, the components interaction preferences are programmable. The self-assembly processes can be easily monitored and analyzed by the components themselves or by external observers (Fig. **11**).

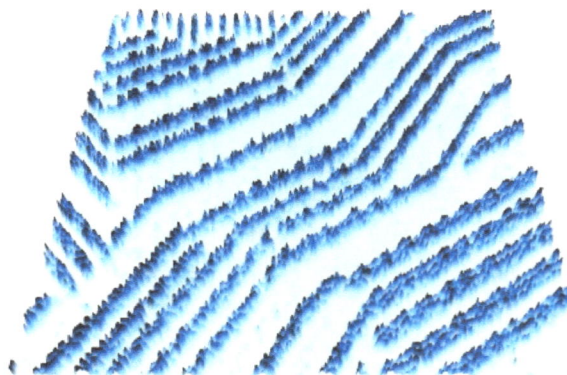

Figure 11: STM image of self-assembled supramolecular chains of the organic semiconductor quinacridone on graphite.

Self-organization and self-assembly are regularly used interchangeably. As complex system science becomes more popular though, there is a higher need to clearly distinguish the differences between the two mechanisms to understand their significance in physical and biological systems. Self-organization is a non-equilibrium process where self-assembly is a spontaneous process that leads toward equilibrium. Self-assembly requires components to remain essentially unchanged throughout the process. Besides the thermodynamic difference between the two, there is also a difference in formation. The first difference is what "encodes the global order of the whole" in self-assembly whereas in self-organization these initial encodings are not necessary. Another slight contrast refers to the minimum number of units needed to make an order. Self-organization appears to have a minimum number of units whereas self-assembly does not.

2. DRIVERS FOR MICRO AND NANOMACHINING TECHNOLOGY

The ability to produce manufactured components to ever more demanding levels of accuracy has been identified by numerous researchers. In 1974, the Japanese precision engineering researcher, Taniguchi, defined *nanotechnology* as the production technology to achieve high accuracy and ultra-fine dimensions of the order of a nanometer [8]. His charts recorded and proposed improvements in machining accuracy capability. These charts can be compared with *Moore's law*, the mid-1960s prediction based on contemporary trends that the integrated circuit (IC) transistor count-per-chip would double every 2 years; it is expected to hold more or less true until at least 2020. Taniguchi stated that the production technology required for nanotechnology must include machine systems for processing, together with appropriate measurement and control techniques. He highlighted that enabling of effective measurement was pivotal to the development of new manufacturing processes and practices. It is probable that a further development of the machining processes can be achieved by extrapolation, in both the microtechnology and nanotechnology regions. It is seen that what was considered as ultra-high-precision machining, for example, in 1928 in the developed countries is considered as normal machining in the same countries in 2000. It is certain that the need for all four classes will continue. The limit for nanoprocessing will be set by the laws of science and, it is probable that this curve will get saturated in the next decade, and the other classes of machining will move parallel to this saturation curve.

It can be estimated that in the early years of the 21st century, the attainable processing accuracy conforms to the nanometer level. Taking the year 2000 as an example, only highly developed countries had the capacity to use the four classes at the tolerances indicated in the figure due to the possibility of attaining the processing accuracy, particularly in the nano-range. The accuracy of processing is expressed by the sum of the systematic error and the random error (3σ standard variance). Systematic errors mainly indicate the failure of a machining system, such as the zero-setting error for tool positioning. Random errors are caused by inherent defects of the processing equipment, such as the presence of a backlash between mechanical links, gears, threads, or sliding guides. Although systematic errors can be corrected by the feedback control of the tool position, random errors cannot. The scattering errors of the machined products or positioning of the tools therefore limit the accuracy of machine tools.

There several reasons for continuously improving the machining precision, namely, improving the quality and property of products, enhancing the stability and reliability of products, minimizing the product, strengthen the interchangeability of parts, improving the efficiency of assembly and promote automotive of assembly. The MNT plays an important role in the manufacturing highly sophisticated products and advanced weapons. Hit accuracy is one of the most important parameters of missile while the hit accuracy depends on the precision of inertial instruments. The ultraprecision machining technology is needed to manufacture the inertial instruments. The precision of gyroscope in the militia III intercontinental ballistic missile is 0.03~0.05°/h which results in the circular probability error of hit accuracy about 500 m. The precision of gyroscope in MX strategic missile (equipped with ten nuclear warheads) is improved more than one magnitude comparing with that of the militia III which ensured the circular probability error of hit accuracy lies in the range of 50~150 meters. It is the basic requirements of many parts of inertial instruments that the manufacturing precision should less equal or less than micrometer level. The gyroscope rotor which weighs one kilogram can result in one hundred meter range error and fifty meter orbit error. The flatness of plane reflector in the laser gyro should range from 0.03 to 0.06 μm, its surface roughness should less than 0.012 μm, its reflectivity should greater than 99.8 percent. The instrument bearing in the manmade satellite belongs to vacuum and no oil lubrication bearing whose surface roughness (R_a) about its bore and shaft should less than one angstrom, its circularity and cylindricity should less than one angstrom. The waveguide is the key component of radar whose property greatly depends on its inner surface roughness which can attain to 0.01~0.02 μm if manufactured by means of MNT, the surface flatness less than 0.1μm, perpendicularity less than 0.1μm, all of these factors ensure quality factor of 6000 which is 2~3 times of what manufactured by traditional method. The reflector which receives infrared ray of the infrared detector is one of the important components of infrared missile, whose surface roughness (R_a) should less than 0.015 μm which can only be satisfied using MNT. The first mirror of Hubble Space Telescope (HST) is a large sized reflector whose diameter attains to 2.4 meter and weighs to 900 kilogram, which makes it owning high resolution ratio. The HST program facilitates the development of six-axis CNC polishing machine be applicable to ultraprecision manufacturing of optical glass thus improving the precision of machining hard-brittle materials. The above facts justify that it is only MNT which can manufacture precision gyroscope, precision radar and other highly sophisticated products.

It is verified by the experiment data of Rolls-Royce corporation that the compression efficiency of automobile engine can be improved from 89% to 94% if the surface roughness decreased from 0.5 μm to 0.2 μm accompanying the improving of manufacturing precision from 60 μm to 20 μm. The torque transmitted by gear box can be increased by one time if the tooth form error of the drive gear can be decreased from 6 μm to 1 μm. The developing of large scale integrated circuit promotes microengineering and also depends on the progress of microengineering. The advancement of integrated circuit request the miniaturization of component which makes more electronic elements can be assembled in the limited areas thus generate complicated circuit. The desire to place many transistors on to a silicon wafer has demanded innovative ways to fabricate electronic circuits and to fit more and more electronic devices into a smaller workable area.

The storage of hard disk largely depends on the distance between magnetic head and disk (namely, the flying height). It is also sometimes called the head gap, and some hard disk manufacturers refer to the heads as riding on an "air bearing". Modern hard disk has a floating height of an amazing 0.5 micro-inch while a human hair has a thickness of over 2,000 micro-inch. The ultra-flatness and super-smooth disk substrate and coatings is necessary to meet this requirements. The memory density of hard disk has already been increased 10,000 times from 1957 to 1982. This progress should largely attribute to the improvement of manufacturing precision and decreasing of surface roughness.

The application scope of precision manufacturing especially MNT is very limited in the past years. Recently, the MNT has already entered each regime of people's everyday life, the application style has transformed from single piece or small batch production to the mass production of reliable and economical nanoscale materials, structures, devices, and systems. Although miniature devices may be manufactured by various procedures, their shaping through removal of material constitutes a major means of production. Established and recently developed methods of machining continue to be investigated for the shaping of such parts to specified small dimensions. The efficiency and the cost of MNT became an important issue as the mass production come into reality.

Presently, the research about MNT mainly centralized in three aspects, namely:

(1) Ultraprecision turning such as single point diamond turning (SPDT), which can machine several kinds of mirror surface and successfully solve the difficulty about machining high accuracy gyroscope, laser reflect mirror and other large scale reflect mirrors.

(2) Precision and ultraprecision grinding, which can be applied to the machining of the substrate of ultra-large scale integrated circuit and the high accuracy hard disk.

(3) Ultraprecision non-traditional machining such as electron beam and ion beam, which advanced the realization of 0.1 μm line-width.

Today, the MNT has achieved great progress and no longer been an isolate machining methodology or pure machining process but has already became a complex system engineering which includes many components. The application of MNT depend on many factors such as ultraprecision machining tool, stable working conditions, the real-time inspection and feedback control using computer technology and so on. It is only through integrating all of corresponding achievements that people can make MNT into reality.

3. THE COMPLEXITY OF MICRO AND NANOMACHINING TECHNOLOGY

3.1. Size Effect in the Micro & Nanomachining

When material is removed by machining there is substantial increase in the specific energy required with decrease in chip size. It is generally believed this is due to the fact that all metals contain defects (grain boundaries, missing and impurity atoms, *etc.*), and when the size of the material removed decreases, the probability of encountering a stress-reducing defect decreases. Since the shear stress and strain in metal cutting is unusually high, discontinuous micro-cracks usually form on the metal-cutting shear plane. If the material being cut is very brittle, or the compressive stress on the shear plane is relatively low, micro-cracks grow into gross cracks giving rise to discontinuous chip formation. When discontinuous micro-cracks form on the shear plane they weld and reform as strain proceeds, thus joining the transport of dislocations in accounting for the total slip of the shear plane. Relatively recently, an alternative explanation for the size effect in cutting was provided based on the premise that shear stress increases with increase in strain rate. It is assumed in the analysis that the Von Mises criterion pertains to the shear plane which is inconsistent with the experimental findings of Merchant. Until this difficulty is taken care of, together with the promised experimental verification of the strain rate approach, it should be assumed that the strain rate effect may be responsible for some notion of the size effect in metal cutting. However, based on the many experiments discussed here, it is very unlikely that it is totally responsible for the size effect in metal cutting.

It is important to understand the basic mechanisms especially when trying to model and predict surface integrity under the machining situations at different length scales. Normally the basic mechanisms are combined in a very complex way due to more complicated contact geometry, materials failure mechanism, due to inhomogeneous surface materials with changing properties and due to variations in loading and sliding conditions. During the sliding contact some of the parameters will change, surface layers are formed, strain hardening takes place, local temperature rise causing softening, *etc.* and after one sliding event we may have a new set of parameters controlling the materials failure and friction.

The main difference between general machining and the micro & nanomachining is the depth of cut. The consequence of changing the depth of cut is to change the scale at which material is removed from the surface. Material removal during brittle machining occurs *via* crack propagation and brittle fracture; and the abraded material is on the order of microns in size. During ductile machining, material removal is the result of plastic fracture and the abraded material is on the order of nanometers in size. An additional feature of ductile grinding is that the surface experiences plastic flow and densification. During micro & nanomachining removal is the result of bond severing on the atomic or molecular scale which is assisted by chemical reactions. Material removal during micro & nanomachining is on the order of atoms and molecules or clusters of atoms and molecules. There are exists complex multilevel relation chain between microscopic molecular behavior and macroscopic surface generation process. The strong coupling among different length scales result in the complexity of surface generation by means of micro & nanomachining. People ought to grasp the overall characteristic and breakthrough the limit of single length scale such as classical mechanics or molecular dynamics which can uncover the physical essence of micro & nanomachining. So, it is useful to investigation the precision machining technique on different size levels using multiscale method. Presently, the number of influencing parameters is large but the situation is still not hopeless to control. In each contact situation (cutting tool & workpiece) there are typically a limited number of some five to ten parameters that dominate the friction and wear behavior. If we can identify them and understand their interactions then we are well on the road to predicting and controlling final surface integrity by means of technique control. Dominating parameters in the micro & nanomachining are the relative speed at the interface, depth of cut, surface roughness and debris in the contact.

The whole picture is becoming even more complicated since we have friction and wear related phenomena involving in the micro & nanomachining process. In some cases we have shearing taking place on a nano level due to molecular or atomic interactions. In other cases we talk about cracks appearing at asperity collisions on a micro level. Or when observing the prevention of contacts by thin film lubrication (cutting fluid) we calculate the pressure and lubricant film thickness on a macro level. Forces and vibrations are observed on component level while the efficiency and lifetime is estimated on machinery level. These length scales of micro & nanomachining represent different approaches to identify and understand characteristic tribology related phenomena.

Here we can talk about micro & nanomachining on two different length scale levels:

Nanomachining includes phenomena related to the interaction between molecules and atoms, such as the effects of van der Waals forces and related interatomic phenomena, determined by the crystal and bonding structures of materials.

Micromachining relates to aspects typically taking place in the bulk materials or at the peaks of the surface topography. Phenomena such as adhesion between asperities, fracture, elastic and plastic deformation, debris formation, surface layer formation and topography changes are all important at this scale.

Origins and phenomena of materials fracture in the machining are manifold. The reason for this can be found in the fact that the phenomena are predominantly determined by the microscopic properties of a material which in turn vary extensively from material to material. In this eBook, emphasis is placed on multiscale coupling continuum-mechanical and microscopic analysis of materials fracture behavior and surface generation.

3.2. Energy Dissipation in the Micro & Nanomachining

Energy dissipation during materials machining processes has captured the attention of engineers and scientists for over 100 years. Why then do we know so little about dissipation in machining processes? A simple answer is that we cannot see what is taking place at the interface during sliding and in the materials while deformation or crack proceeds at different levels. Recently, however, devices such as the atomic force microscope have been used to perform friction measurements, characterize contact conditions and even describe the "worn surface". Following these and other experimental developments, friction modeling at the atomic level-particularly molecular dynamics simulations - has brought scientists a step closer to "seeing" what takes place during machining. With these investigations have come some answers and new questions about the modes and mechanisms of energy dissipation at the sliding interface or materials internal. This eBook will review recent results of 1) molecular dynamics and other theoretical studies that have identified modes of energy dissipation during machining processes and 2) continuum mechanics based finite element investigation which added to our understanding of dissipation processes at macroscopic scale. Finally, several approaches for addressing the questions of dissipation mechanisms will be presented.

The energy which supplied by means of basic particles such as atom, molecule and electron from region with high temperature to the region with low temperature is termed as thermal energy. The ways by which the energy flowing is closely related with energy dissipation. Presently, there are basically three kinds of heat loss method: conduction, convection and radiation. In reality, the heat loss process may be involved one or several kinds of above means. In the micro & nanomachining process, the non-uniform temperature distribution is the result of poor heat conduction while bring down the temperature at the interface ought to be contributed to the heat convection realized by cutting fluids.

3.2.1. Conduction

Conduction is accomplished through the two following methods:

(1) Interaction among molecules, the high-energy particles collide with low-energy particles around themselves intensely and pass the energy to low-energy particles;

(2) The effect realized by free electron or phonon, the heat transfer by free electron mainly generated in pure solid while in non-metal crystal this is realized by phonon; in the alloy both of free electron and phonon attend in the heat transfer process.

The heat conduction is a kind of molecule based physical process, the heat transfer model is similar with the model of molecular momentum transfer, which firstly brought forward by *Fourier* at 1892 as follows:

$$q_x = -\lambda \frac{d\theta}{dx} \tag{1.1}$$

where q_x denotes the density of heat flow, λ is coefficients of heat conduction which justifies the materials ability of heat conduction. The negative signal in the equation illustrate that the direction of heat conduction is opposite to that of heat gradient direction, namely, the heat conduction will forever along the direction of dropping of temperature.

3.2.2. Convection

Convection implies the heat transfer induced by the macroscopic relative motion of different component with different temperature in the fluid. The most common case on heat transfer in the engineering is the between the fluid and the solid wall with relative motion which is also termed as convection heat transfer. Therefore, the motion of the fluid has a great influence upon the efficiency of the convection. The heat transfer equation is bright forward by *Issac Newton* which also named as Newton heat transfer model:

$$q = \alpha(T_w - T_f) \tag{1.2}$$

where q is heat density, T_w is the temperature of solid wall, T_f is the stable temperature of fluid adjacent to the wall, α is the heat transfer coefficient which is the function of geometry shape, fluid property and the temperature difference.

Surface interactions at tribological interfaces (tool rake surface & chip; tool flake surface & machined surface) in the micro & nanomachining process are highly complex, and their understanding requires knowledge of various disciplines including physics, chemistry, applied mathematics, solid mechanics, fluid mechanics, thermodynamics, heat transfer, materials science, rheology, lubrication, machine design, performance and reliability. Understanding the tribological interactions discussed can only be possible if the energy involved is known, because any interaction should be considered as some kind of energy exchange. The amount of energy transmitted through tribological interface defines to a large extent the actual occurrence of various physical and chemical processes that might happen at this interface because any of these processes requires a certain level of energy to trigger and maintain this process.

The shearing, cutting and rubbing processes involved in machining generate large heat flow, which can only be removed to a small extent with the chips and thus can lead to considerable thermal stress on the workpiece and the tool. Analysis the heat generation and loss in the machining is of great importance to surface integrity. As the depth of cut decreases to nanometer scale, most of the cutting operation is accomplished by the local tool tip. There are different controlling mechanisms dominate the materials removal and tribology behavior which may change the heat generation. Furthermore, the working conditions also changed greatly as the different contact conditions at the workpiece-tool interface which take an effect upon heat loss. Materials removal in the micro & nanomachining is on the atoms or molecules or cluster of atoms or molecules. The origins of dissipation in ultraprecision machining are related to phonon excitation, electronic excitation and irreversible changes of the surface. In a typical micro & nanomachining experiment, the energy dissipated in a single atomic slip event is of the order of 10 eV. In materials deformation/removal process at nanometer scale, power dissipation caused by internal/external friction is observed, which is proportional to the amount of materials to be removed and materials fracture strength at this scale.

In micro & nanomachining, most of generated heat is absorbed by the cooling lubricant. Under this circumstance, the lubricant holds two primary functions:

- reducing the friction between the tool and the workpiece by forming a stable lubrication film and

- cooling the contact zone and the workpiece surface by absorbing and transporting heat.

Specific secondary functions include:

- purifying the cutting tool and the workpiece,

- chip transport from the machining location and

- building up corrosion resistance for both the cutting tool and the workpiece material

When estimating the contact area in the contact mode for a few atoms, the energy dissipation per atom that can be associated with a bond being broken and reformed is also around 100 meV. The idea of relating the additional damping of the tip oscillation to dissipative tip-sample interactions has recently attracted much attention. The origins of this additional dissipation are manifold: one may distinguish between apparent energy dissipation (for example from inharmonic cantilever motion, artefacts from the phase controller, or slow fluctuations round the steady state solution), velocity-dependent dissipation (for example electric and magnetic field-mediated Joule dissipation) and hysteresis-related dissipation (due to atomic instabilities or hysteresis due to adhesion).

3.3. Fractal Analysis of Surface Roughness

Many modern high techniques depend for their appropriate function on the bulk properties, but for an important group of phenomena these properties are the surface properties. The behavior of components therefore greatly depends on the surface of the material, surface contact area and environment under which the material operate. Engineering surfaces generated by micro & nanomanufacturing technology which are left with the texture caused by the major operation formed the bulk shape. In general, surfaces are designed to interact with other solids which lead to the surface preparation is often aimed at producing functional surface. It is valuable to develop surface parameters as a means of communication between design, manufacture and quality control. In traditional approaches, we usually postulated surface morphology as a stationary random process. Thus, statistical parameters such as variance of height, slope and curvature are used. However, it has been observed that surface topography is a non-stationary random process, implying that the statistical parameters are not always independent of length scales, and are related to sampling length and resolution of the measuring instrument for a particular surface. Many statistical parameters are not intrinsic parameters of the surface, and in order to better characterize the surface, the only method is to enlarge the number of parameters. As a result, such a large number of characterization parameters occur that the term "parameters rash" is aptly used. Therefore it is very important to characterize rough surfaces using intrinsic parameters which are independent of the sampling length or area.

Fractal theory is a good choice for the characterization of surface morphology. Fractal theory was first proposed by Benoit Mandelbort in the study of "How long is the coast of Britain" [9]. After that, it is confirmed that many natural things and phenomena, such as topographic heights, rainfall and copper grades, appeared fractal, and estimated the fractal dimension of various environment series. Since then, fractal theory has been widely used in geomorphology, physics, finance and other fields. In the field of surface engineering, fractal theory was used to study contact mechanics, wear, cracks analysis and characterization of surface morphology, and bear developed some achievements accordingly. However, there are still some problems. For instance, different scholars reached the opposite conclusion when they study the relationship between fracture toughness and fractal dimension.

Fractal geometry initially used to describe the irregular geometry. It provides a more scientific approach for the characterization of rough surface. Some scholars found that most surfaces obey, at least in some scale regime, the fractal self-affine symmetry. W-M function is widely used in characterization of engineering surfaces, and it has solid theoretical basis.

4. ORGANIZATION OF THIS EBOOK

It is seen nowadays that emphasis is on manufacturing high-precision products cheaply and quickly. The history of increasing machining precision also suggests that there is an ever-increasing demand for creating value-added products. The manufacture of high performance high-value-added computers was made possible because of the progress in machining technology. Thus, micro & nanomachining is the science of creating, directly and indirectly, greater-value-added products in various fields, which form the foundations of our modern advanced civilization. The need to manufacture high precision items and to machine difficult-to-cut materials led to the development of the newer machining processes. The dimensional tolerance achieved by precision machining technology is on the order of 0.01 μm and the surface roughness is on the order of 1 nm. The dimensions of the parts or elements of the parts produced may be as small as 1 μm, and the resolution and the repeatability of the machine used must be of the order of 0.01 μm (10 nm). The accuracy targets for ultraprecision component cannot be achieved by a simple extension of conventional machining processes and techniques. They are called micro & nanomachining processes, notwithstanding that the definition of conventional and traditional changes with time. Unlike conventional machining processes, micro & nanomachining processes are not based on the removing the metal in the form of chips using a wedge shaped tool. There are a variety of ways by which the material may be removed in precision machining processes. Some of them are abrasion by abrasive particles, impact of water, thermal action, chemical action and so on. The challenges to micro & nanomachining are many, beginning with the incredibly broad range of applications, materials, and geometries that have been proposed for nanoscale structures. However, components made from engineering materials require shaping processes other than those established for general purpose machining technique. Therefore, traditional machining processes require further development in order to machine components that are fit for purpose at the micro and nanoscales. The range of micro & nanomachining routes towards the structures is almost as diverse as the materials, applications, and geometries needed for next-generation applications. The approaches can begin to be compartmentalized by separation into either a "top-down" or "bottom-up" approach. In many cases, these micro & nano machining advances have produced stunning physical and chemical discoveries and phenomena that have no analog in larger scale structures. Scaling laws governing the mechanical behavior of materials from atomistic (nano), *via* mesoplastic (micro), to continuum (macro) scales are very important to numerous applications, such as the development of a new class of aircraft engine materials, or new steels for naval battle ships, or new tank armor materials for the army, or numerous microelectromechanical components for a myriad of applications. Scaling laws are also important for applications where two length scales of different orders of magnitude are involved. For example, one is atomistic (nano) and the other mesoplastic (micro) as in nanoindentation, or, one is mesoplastic (micro) and the other continuum (macro) as in conventional indentation. Appropriate scaling laws may extend the extensive knowledge accumulated over time on material behavior at the macro (or continuum) level to the atomistic (or nano) level, *via* mesoplastic (micro) level. Therefore, some of the newest advancements about MNT are provided in chapter 1.

In the micro & nanomachining process, materials undergoes severe plastic deformation in passing through a highly localized shear zone extending from the tip of the tool to the free surface at the juncture of chip and undeformed workpiece. When the cutting tool moves through the workpiece material, the chip is separated from the workpiece and slides over the rake face of the tool. The highly pressurized friction between the chip and rake face causes the further straining of the materials in the area around tool surface. The plastic deformation and friction generate large amount of thermal energy in the cutting zones which raises the temperature high enough and sometimes beyond the re-crystallization temperature of the materials. The process variables depend upon the cutting conditions and tool geometry which bring about large variety to the process. A successful cutting model ought to be used to incorporate such effects. In fact, many technical parameters such as strain, strain rate and temperature are difficult to be measured and direct data about material at conditions similar to what occurs in machining is hardly available. Mechanism of surface generation and chip formation and many new phenomena happened in the machining process still lacks meaningful fundamental research. Numerical modeling of micro & nanomachining technique can provide insight into the mechanics of surface generation and chip formation and predict the process variables with reasonable accuracy and cost. This eBook provides some meaningful research on MNT using molecular dynamics (chapter 2) and finite element method (chapter 3) for phenomena happened in different scale. There is a prospect that one day, such simulations may replace costly experiments for fundamental study, tool design and process planning.

The study and development of micro & nanomachining technology has been an area of ongoing focus for numerous researchers. Interest in this topic has been increased over the past decade due to the trend towards higher accuracy, smaller sized components. Linking material property acquisition and modeling in the nanometric scale with those on the micro-scale is a considerable challenge in material science in general and in micromachining in particular. The detailed knowledge of material behavior will provide the necessary insight to support process development, modeling and the optimization of critical ultraprecision machining processes. Multiscale methods offer the best hope for bridging the gap that exists between different scales for studying and understanding the behavior of materials in the machining process. Here the "multiscale" broadly means that it involves phenomena at disparate length and/or time scales spanning several orders of magnitude. Understanding of the fundamental physical mechanisms behind micro & nanomachining is a major scientific task and it is the reason why multiscale analysis is so important. This is based on the judgment that multiscale investigates the effects of a hierarchy of internal structures on material behavior, thus making it realistic to understand fundamental mechanisms and to avoid misunderstandings based on macroscopic observations and resulting assumptions. Multiscale method conform to the basic philosophy ideas, namely, every things should evolve from quantitative change to qualitative change. Owing to the central role that multiple scale methodology appears poised to play in the manufacturing science in the foreseeable future, the author carries out some meaningful research about physical essence of micro & nanomachining based upon multiscale method which can bridge the physical essence and engineering applications (chapter 4).

MNT is an extremely complex phenomenon ought to be studied in depth using ideas of modern physics that deal with the problems of complexity. The fractal surface, non-equilibrium thermodynamics, self-organization process in the complex machining process has already beyond the general mechanical study regime. Some new theoretical tools are needed to uncover the physical essence of MNT. Therefore, some new investigations based on complexity are introduced in chapter 5.

This eBook is developed to meet the growing need of mechanical engineers, and others, to understand the design and process issues associated with precision machine tools and the fabrication of precision components. Furthermore, this eBook also tried to introduce some fundamental understanding of complex mechanism and multiscale analysis method. These are big topics and we don't claim to cover all to some sufficient depth in this eBook. But, the foundations laid here can be built upon for additional study. A remarkable enhancement in computational capability (computer hardware) and high performance computation techniques (parallel computation) has enable researcher to employ the large scale parallel numerical simulation to investigate the micro & nanomachining and gain insights into this process.

REFERENCES

[1]   P.A. McKeown, "High precision manufacturing in an advanced industrial economy," James Clayton Memorial Lecture, Inst. Mech. Eng., London, 23rd April 1986.
[2]   N.B. Dahotre, S.P. Harimkar, *Laser fabrication and machining of materials*. New York: Springer, 2008
[3]   M.J. Jackson, *Micro and nanomanufacturing*. New York: Springer, 2007
[4]   G. Carter, J. S. Colligon, *Ion Bombardment of Solids*. London: Heinemann, 1968.
[5]   J. McGeough, *Micromachining of engineering materials*. New York: Marcel Dekker, Inc., 2002
[6]   K. H. Steigerwald, E. Meyer, "New developments in electron beam machining methods, electrical methods of machining and forming. Institution of Electrical Engineers", Conf. Publ., no. 38, pp. 252-258, August 1967.

[7]   D. Boehme, "Perforation welding and surface treatment with electron and laser beam", Proc. 7th Int. Symp. on Electromachining IFS, no. 189,pp. 200-207, June 1983.

[8]   N. Taniguchi, "Current Status in and future of ultra-precision machining and ultrafine materials processing", *Annals CIRP*, vol. 32(2), pp. 573-582, February 1983.

[9]   B.B. Mandelbrot, "How long is the coast of Britain? Statistical self-similarity and fractional dimension", *Science*, vol. 156, pp. 636-638, June 1967.

Send Orders for Reprints to reprints@benthamscience.net
Micro and Nanomachining Technology Size, Model and Complex Mechanism, 2014, 49-123   **49**

CHAPTER 2

Molecular Dynamics Simulation of MNT

Abstract: The machining allowance and chip thickness can be reduced to less than approximately 1 µm in ultraprecision manufacturing such as single point diamond turning (SPDT). Metal surfaces can be finished under precisely controlled machine and environmental conditions. Accuracies of, respectively, 10 and 1 nm have been attained in practice and under experimental conditions (aided by advanced control techniques). Under the highly precise motion of such a machine tool, the primary factor affecting machining accuracy is the controllability or repeatability of the thickness of cut, that is, the undeformed chip thickness effectively removed at the cutting edge. Experiments with a specially prepared fine diamond cutting tool used on such a machine tool have confirmed that very fine chips, of undeformed thickness as small as 1 nm, can be removed in the turning of some highly machineable work materials. However, the accuracy attainable, and mechanisms such as chip formation and surface generation in microcutting are still not well understood, owing to the limitations in availability of experimental and measurement techniques, and in analytic methods for studying such machining conditions. Micro & nanocutting that occurs in a small region which contains only a few layers of molecules can consequently be atomistic, or discrete in nature, rather than continuous, as is assumed in conventional continuum mechanics. In studies of such atomistic processes, which are difficult to investigate experimentally, computer simulation by molecular techniques is useful. Further advancements in the machining technology can be aided through a theoretical understanding of micro & nanomachining. Molecular dynamics (MD) simulation, like other simulation techniques can play a significant role in addressing a number of machining problems at the atomic scale. It may be noted that atomic simulations are providing new data and exciting insights into various phenomenon in micro & nanomanufacturing processes that cannot be obtained readily by any other theory or experiment. The diversity of these processes renders difficult of using a generalized theoretical analysis of micro & nanomachining, however, the methods of molecular dynamics are becoming increasingly attractive for studies of micromachining, especially as the technology advances toward the shaping of parts in the nanometric range. The foundations of molecular dynamics that are needed for theoretical treatments of micromachining therefore form the basis of MNT. Several such analyzing of micro & nanomachining have been developed. In this chapter, the principle of molecular dynamics (MD) simulation on micro & nanomachining and the procedures used to determine the accuracies attainable are described. As noted above, the case of diamond machining and chemical mechanical polishing are used to illustrate the technique, although molecular dynamics is now being increasingly used in studies of many other methods of micro & nanomachining.

Keywords: Molecular dynamics, statistical ensembles, finite differential methods, interatomic potential functions, geometrical-physical model.

Xuesong Han

INTRODUCTION

Machining allowances in conventional metal cutting can be as high as hundreds of μm. However, in ultraprecision cutting, such as single point diamond turning (SPDT), the machining allowance and chip thickness can be reduced to less than approximately tens of micrometer or nanometer. Metal surfaces can be finished under precisely controlled machine and environmental conditions. Accuracies of, respectively, 10 and 1 nm have been attained in practice and under experimental conditions (aided by advanced control techniques). Under the highly precise motion of such a machine tool, the primary factor affecting machining accuracy is the controllability or repeatability of the thickness of cut, that is, the undeformed chip thickness effectively removed at the cutting edge. Experiments with a specially prepared fine diamond cutting tool used on such a machine tool have confirmed that very fine chips, of undeformed thickness as small as 1 nm, can be removed in the turning of some highly machineable work materials. However, the accuracy attainable, and mechanisms such as chip formation and surface generation in microcutting are still not well understood, owing to limitations in availability of experimental and measurement techniques, and in analytic methods for studying such machining conditions. Furthermore, microcutting that occurs in a small region which contains only a few layers of molecules can consequently be atomistic, or discrete in nature, rather than continuous, as is assumed in conventional continuum mechanics. In studies of such atomistic processes, which are difficult to investigate experimentally, computer simulation by molecular techniques is useful. Several such analysis of microcutting have been developed. Over the last decades, there has been a new realization that understanding the nanoscale behavior is required for understanding how materials fail (removal), which opens great opportunities to provide new methods for materials processing from the bottom up. The significance of atomistic processes to describe deformation and failure of materials is apparent for understanding physical essence of MNT.

Molecular dynamics (MD) simulation is introduced to model materials property and nanometric cutting since thirty years ago. Unlike the finite element method, in

MD simulation nodes and the distance between nodes are selected not on an arbitrary basis but on more fundamental units of the material, namely, center of the atom as the nodes which enables the process can be reduced to its fundamental units for analysis. Also, MD techniques give higher temporal and spatial resolution of the cutting process than is possible by other numerical computation approach. Consequently, certain phenomena of necessity neglected in continuum analysis can be effectively investigated by MD simulation. However, since a large number of atoms constitute any common material, it is necessary to consider the interactions of several thousands of atoms in MD simulation of machining. Since such a simulation requires large mainframe computers with significant memory and fast processing times, this research was mostly confined to the few national laboratories until recently such as the Lawrence Livermore National Laboratories (LLNL) of USA. Although universities have access to supercomputers located at various regions in the world, time sharing by multiple users limited its use for this application. Recent availability of fast workstations with significant memory (almost comparable with mainframe computers) has now enabled research to be conducted in this area at academic institutions. If the results of the MD simulations are to be useful in the interpretation and prediction of experimental data, appropriate care must be exercised in principles of molecular dynamics.

This chapter covers computational methods and techniques operating at the atomic scale, and describes how these techniques can be used to model the dynamics of materials removal and other deformation mechanisms. A description of molecular dynamics as a numerical modeling tool covers the use of interatomic potentials (pair potentials such as the Lennard-Jones model, embedded atom method (EAM), bond order potentials such as Tersoff's and Brenner's force fields) in addition to the general philosophies of model building, simulation, interpretation, and analysis of simulation results. Example applications for specific materials (such as silicon, copper) are provided as case studies for MNT, and problems discussed. Readers will find a physics-motivated discussion of the numerical techniques along with a review of mathematical concepts and code implementation issues. Using specific examples such as investigations of crack dynamics in brittle materials or deformation mechanics of nanomaterials, this volume conveys how atomistic studies have helped to advance developing new theories, or provided insight into the molecular deformation mechanisms, explaining or supplementing experimental results. Many of the examples are adapted from studies carried out by the author of this eBook, and some of the discussion should therefore not be considered as a comprehensive and inclusive review with respect to the wider range of available results. Rather, they represent a set of specific examples to illustrate the application of the atomistic simulation techniques reviewed here. This chapter mainly presents an introduction into molecular dynamics simulation approaches and the corresponding discussion includes the physical basics, numerical implementation and examples of atomistic models for specific applications.

1. BASIC CONCEPTS OF MOLECULAR DYNAMICS

Some problems in the statistical mechanics can be exactly soluble, that is to say, complete specification of the microscopic properties of system can be used to describe materials macroscopic properties. But in reality there are only a few non-trivial problems in the statistical mechanics can be solved. In fact, more problems cannot be solved exactly can be studied using straightforward approximation scheme. Computer simulation can plays an important role in providing essentially exact results in many complex systems which cannot be solved by other method. Computer simulation can provide direct route from the microscopic details of a system to macroscopic properties of experimental interests. As in many micro & nanomachining technology, the precision surface is always acquired by means of materials removal in the small unit such as atom by atom or molecular by molecular. It is difficult to carry out experiments on subtle details of dynamic molecular motion and the structure. Molecular dynamics [1-2] simulation is playing an increasingly prominent role in the analysis of the behavior of materials at an atomistic level that cannot be readily obtained either by other theoretical methods or by experiments. It is a methodology for investigating statistical properties of condensed matter systems. Predictions based on an atomistic level of understanding are providing increasingly useful and accurate information for a myriad of applications in material science, tribology and machining. Theoretical description of complex systems based on statistical physics has already been well developed but can be complex to implement while computer simulation may provide an efficient solution to study the specific aspects of a system in great detail. Molecular dynamics method was first introduced by Alder and Wainwright [3] in the late 1950's to study the interaction of hard sphere. After that, many important insights concerning the behavior of simple liquids emerged from their studies. In 1964, Rahaman [4] carried out the first simulation using a realistic potential for liquid argon which may be considered as the starting point for dynamic calculations. It was the first work where an exact method was used to calculate dynamic quantities and transport coefficients for a real system. The first molecular dynamics simulations of real system were carried out by Rahman and Stilinger [5] in their modeling of liquid water in 1974. Since then, the number of simulation effort has greatly expanded.

Advancements in machining technique cannot be achieved by simply "fine turning" process. There is a need for the development of fundamental understanding of the many facts of MNT in order to improve the efficiency and precision. Molecular dynamics simulation is a numerical technique for computing the equilibrium and transport properties of classical many-particle system which can provide a fundamental description of the materials fracture (removal) and deformation happened at atomic level in MNT. In such cases one must turn to the microscopic basis of matter and design a theory based on the molecular properties of a solid that requires only few data or is even fully predictive. Over the last decades, there has been a new realization that understanding the nanoscale behavior is required for understanding how materials removal. We also give a first introduction to the primary concepts of the microscopic world, including a brief glance at the properties of real molecules and the philosophy behind formulating molecular models. Assuming an atomistic viewpoint has another quite important aspect: It allows for the seamless communication the study of deformation and fracture. The reason is that the notion of the "chemical bond" is a concept with which many disciplines can be linked. Atomistic models typically contain extremely large number of particles, even though the actual physical dimension may be quite small. For example, even a crystal with dimensions below a few micrometers side length has several tens of billions of atoms. Predicting the behavior of such large many particle systems under explicit consideration of the trajectory of each particle is only possibly by numerical simulation, and must typically involve large computational facilities.

Furthermore, the significance of including atomistic or nanoscale mechanisms is also due to the increasing trend toward miniaturization. Once the dimensions of materials reach the submicron length scale, the continuum description of materials is questionable and the full atomistic information about the material state is often necessary to study relevant materials phenomena. The ability to create nanostructures by machining has reached a level of perfection that now enables us to make almost arbitrary structures and shape at ultra-small scales. However, these advances also require new theoretical concepts to describe the materials behavior, as it can be quite different from what is conventionally known from larger scale materials. Atomistic models are often quite suitable to capture these and other effects. The observed phenomena in a larger ensemble of atoms can bridge the atomic scale to microscopic or macroscopic scale. However, since the basic results of an atomistic simulation are only atomic positions, velocities, and forces, the interpretation of these numbers can be quite challenging.

Calculating the partition function and associated thermodynamic and equilibrium properties for a general many-body potential that includes nonlinear interactions becomes an insurmountable task if only analytical techniques are employed. Unless a system can be transformed into a more tractable form, it is very unlikely that the integrals of particle movement equation can be performed analytically. In this case, the only recourses are to introduce simplifying approximations, replace a given system by a simpler model system, or employ numerical methods. Molecular dynamics is a technique that allows a numerical "thought experiment" to be carried out using a model that, to a limited extent, approximates a real physical or chemical system. Such a "virtual laboratory" approach has the advantage that many such "experiments" can be easily set up and carried out in succession by simply varying the control parameters. Moreover, extreme conditions, such as high temperature and pressure, can be created in a straightforward (and considerably safer) manner. The obvious downside is that the results are only as good as the numerical model. In addition, the results can be artificially biased if the molecular dynamics calculation is unable to sample an adequate number of microstates over the time it is allowed to run.

Generally speaking, molecular dynamics is a model based microscopic description of real physical system. The system can be few- or many-body system. The description may be a Hamiltonian or, Lagrangian or expressed directly in Newton equations of motion. The molecular dynamics method calculates properties using the equation of motion and obtains the static as well as the dynamic properties of system.

The essence of the MD simulation method is the numerical solution of Newton's equation of motion for an ensemble of atoms. These equations are integrated by numerical techniques for extremely short time intervals (2-3fs), and equilibrium statistical averages are computed as temporal averages over the observation time. In principle, explicit knowledge of the electronic ground state in each system configuration is required in order to have a correct description of the interatomic forces. To render atomistic simulation studies practical, a classical or semi-classical potential from which interatomic forces can be derived is necessary. This is accomplished by using an appropriate empirical potential energy function that satisfies stringent material properties criteria that include the lattice constant, energy of sublimation, compressibility, elastic constants, the equation of state and the stability of the crystal itself. Experiments are generally performed on a macroscopic sample which contains a larger number of atoms or molecules sampling an enormous number of conformations. In statistical mechanics, averages corresponding to experimentally measured values are defined in terms of ensemble averages. An ensemble average is an average take over a large number of replicas of the system considered simultaneously. In MD simulations, the points in the ensemble are calculated sequentially in time, so in order to calculate the ensemble average, the MD simulation must pass through all possible states corresponding to the particular thermodynamic constraints. In reality, the MD simulation calculates the time average while experimental results correspond to ensemble average. The problem is solved by the assumption that the time average is equal to the ensemble average:

$$<A>_{Ensemble} = <A>_{Time} \tag{2.1}$$

In order for MD to serve a useful purpose it must be capable of sampling a representative region of the total phase space of the system. An obvious corollary of this requirement is that the results of a simulation of adequate duration are insensitive to the initial state, so that any convenient initial state is allowed. A particularly simple choice is to start with the atoms at the sites of a regular lattice - such as the square or simple cubic lattice - spaced to give the desired density. The initial velocities are assigned random directions and a fixed magnitude based on temperature; they are also adjusted to ensure that the center of mass of the system is at rest, thereby eliminating any overall flow. The speed of equilibration to a state in which there is no memory of this arbitrarily selected initial configuration is normally quite rapid, so that more careful attempts at constructing a "typical" state are of little benefit.

Molecular dynamics simulation is meaningful because it is applied to precisely defined model for the material behavior in MNT. Models for molecular dynamics simulation basically composed of two kinds: one for the interactions among molecules (basic particles) making up the system and another for interactions between molecules and the surrounding environmental media. The model for molecular interactions is contained in an intermolecular force law or intermolecular potential energy function. The potential function implicitly describes the geometric shapes of individual molecules or their electron clouds. A detailed characterization of intermolecular potential functions can be given analytically or numerically. In most simulations the intermolecular potential energy is taken to be a sum of isolated pair interactions as shown in the following:

$$U = \sum_{i \neq j} \sum u(r_{ij}) \tag{2.2}$$

here $u(r_{ij})$ is a pair potential energy, r_{ij} is the distance between molecules i and j. As there are no dissipative forces act among molecules, intermolecular forces are conservative therefore the force acted upon the ith atom can be derived as follows:

$$F_i = -\frac{\partial U}{\partial r_i} \tag{2.3}$$

here $\partial / \partial r_i$ represents the gradient operator, where repulsive force is positive and the attractive force is negative. Besides, another important factor about molecular modeling is boundary conditions which define the method of interaction between molecules and their surroundings such as heat transfer. In reality, characteristics of boundary conditions mainly depend upon the physical process to be simulated although some freedom usually exists in the way in which boundary conditions are realized. For example, if bulk fluid is to be analyzed, hard boundaries should be avoided. If non-uniform regions such as fluid-fluid or fluid-solid interfaces are to be simulated, boundary conditions to mimic these situations are needed. If the shear or bulk viscosity is to be determined, we may introduce moving boundaries that shear or compress the system. Setting boundary conditions is another necessary step to complete the definition of model system no matter investigation any physical process. Generally speaking, the simulation of molecular process involves four essential parts: (a) construction of molecular model, (b) calculation of molecular trajectories by means of integration of classical equations, (c) simulation of experimental conditions, (d) analyze the simulation results and obtain the necessary property. In molecular simulations, while atoms vibrate around their minimum-energy positions, the minimum-energy positions move with cutting process. The atomic positions are obtained by numerically solving differential equations of motion and hence it is connected with time-the positions reveal dynamics of individual molecules. But this method is computationally intense. In other simulation methods the molecule positions are not temporally related such as *Monte Carlo* method and *Molecular Statics* method. In *Molecular Statics* method, only the positions at which the resultant force on each atom is zero are followed. Of course, the atoms in this case follow the positions of minimum potential energy. It is thus a quasi-static method. Only a few hundred atoms can be considered in this simulation and hence should result in significantly shorter computational time. However, the interpretation and analysis of data by this simulation are rather complex. In *Monte Carlo* simulations the positions are generated stochastically such that molecular configuration depends only on the previous configuration. When the outcome of a random event in a sequence depends only on the outcome of the immediately previous event, the sequence is called *Markov Chain*. These various methods are briefly described to place molecular dynamics within the wider context of molecular-scale simulation methods.

1.1. Different Ensembles in the Molecular Dynamics Simulation

Molecular dynamics simulation is a theoretical toolkit of studying the natural time evolution of classical system composed of N particles in volume V and converse this very detailed information into macroscopic terms. Researches wish to understand the way in which the relatively complicated microscopic many-body dynamics gives rise to the relatively simple macroscopic few-variable behavior described by phenomenological thermodynamics and hydrodynamics. The thermodynamic state of small scale system is usually defined by a set of parameters (such as the number of particles N, the temperature T and the pressure P). Although there are still many other thermodynamic quantities, all of them can be derived through knowledge of the equations of state and the fundamental equations of thermodynamics. While most of these parameters convey some properties about the microscopic structure and dynamics of the system, their values are completely dictated by the few variables (such as NPT) characterizing the thermodynamic state but not by the very many atomic positions and momenta that define the instantaneous mechanical state. Some of the variables useful for describing macroscopic systems have obvious microscopic analogs. The macroscopic mass ρdr and the momentum $\rho v dr$ in a volume element dr, for instance, correspond to simple sums of one-particle contributions. The macroscopic energy and the pressure tensor are more complicated functions. They include not only one-particle kinetic part, but also two- or more-particle potential contributions.

The entropy and free energy functions are even more complicated than mass, momentum, energy and pressure. For small few-particle volume elements there is no sensible and appealing definition of entropy guaranteed to resemble thermodynamic entropy. But for large volume elements, Gibbs showed that the entropy corresponds to the logarithm of the available phase-space volume. This phase-space volume can only be determined by explicitly carrying out calculations over an interval of time or by integrating over the appropriate phase space. Thus the entropy depends upon the current state of the system in a relatively complicated way. Fluctuations likewise involve either time or phase-space average, and are more complicated evaluate than mass or energy sums. In transient nonequilibrium systems, far from equilibrium, it is not practical to define instantaneous properties in terms of constrained time or phase-space averaging, and are more complicated to evaluate than mass or momentum or energy sums. In transient nonequilibrium systems, far from equilibrium, it is not practical to define instantaneous properties in terms of constrained time or phase-space averages. This is because such systems change with time, so that the variables constraining a "phase-space-average" are not apparent. Nevertheless, no useful description of time-dependent nonequilibrium behavior is possible without some recipe for describing the instantaneous state of a system. Accordingly, we here consider the overall variables describing a microscopic many-body system.

In reality, energy, volume, and the number of particles are the independent variables describing either an isolated system or a microcanonical ensemble of such systems. For any individual closed and isolated system, obeying Newton's, or Lagrange's, or Hamilton's equations of motion, these properties are constants, "constants of the motion." The only other known constant of the motion, for most interparticle force laws, is the momentum. It is convenient to divide the other nonconserved macroscopic variables needed to describe a many-body system into two categories. The first categories include "mechanical variables". The mechanical variables of dynamical systems can be usefully defined as instantaneous functions of the sets of coordinates and velocities or the coordinates and momenta. These mechanical variables include not just mass, momentum and energy but also the fluxes of these quantities. Provided that the interparticle interactions are pairwise-addictive, the mechanical variables most useful to a hydrodynamic description can be expressed in terms of coordinates and momenta by using simple one-body and two-body functions. There is second category of macroscopic variables, involving the thermodynamic entropy which could be termed as "entropic variables". These depend upon *Gibbs*'s statistical definition of thermodynamic state, and include entropy contribution based upon phase-space volume. The thermodynamic entropy is examples taken from this second category.

Molecular simulations permit the rigorous application of the principles of statistical mechanics to calculate the properties of a chosen system. From the viewpoints of statistical mechanics, a system's macroscopic properties such as pressure or density represent averages over all possible quantum states. To calculate the properties of real system, it is convenient to define an ensemble. An ensemble can be regarded as an imaginary collection of a very large number of systems in different quantum states with common macroscopic attributes. For example, each system of the ensemble must have the same temperature, pressure and the number of molecules as the real system it represents. The ensemble average of any property $<A>$ can be obtained from the relationship:

$$< A >= \sum_{i=1}^{n} A_i p_i \tag{2.4}$$

here A_i is the value of A in quantum state i, p_i represents the probability of observing the ith state, and the angled brackets denote an ensemble average. The time-averaged properties of the real system are related to the ensemble average as stated in the previous section. The equivalence of the time-average and the ensemble-average is called the ergodic hypothesis. The form of p_i is determined by which macroscopic properties are common to all of the systems of the ensemble. For example, if the number of molecules N, volume V and temperature T are constant, then it can be shown that

$$p_i = \frac{e^{-\beta E_i(N,V)}}{Z_{NVT}} \tag{2.5}$$

here E_i is the energy, $\beta = 1/kT$ and Z_{NVT} is the partition function which can be acquired as follows:

$$Z_{NVT} = \sum_{i=1}^{n} e^{-\beta E_i(N,V)} \tag{2.6}$$

This equation represents the canonical ensemble with fixed numbers of molecules which proceed to heat exchange with external sources. Other ensembles are possible depending on the choice of constant macroscopic properties. The microcanonical, canonical and isothermal-isobaric ensembles describe closed system for which there is no change in the number of particles. In contrast, the grand canonical ensemble is appropriate for an open system in which the number of particles can change. In the thermodynamic limit, all of the ensembles are equivalent and it is possible to transform between different ensembles. The choice of ensemble for a simulation is entirely a matter of convenience. Many physical processes do not involve a change in the number of molecules and occur at either constant temperature and volume, or constant temperature and pressure. In these cases, the natural choices are the canonical and isothermal-isobaric ensembles, respectively (Table **1**).

Table 1: Common statistical ensembles

Ensemble	Constraints	Z	p_i
Microcanonical	N,V,E	$\sum_i \delta(E_i - E)$	$\dfrac{\delta(E_i - E)}{Z_{NVT}}$
Canonical	N,V,T	$\sum_i e^{-\beta E_i (N,V)}$	$\dfrac{e^{-\beta E_i (N,V)}}{Z_{NVT}}$
Grand Canonical	μ,V,T	$\sum_i e^{\beta N_i \mu} Z_{NVT}$	$\dfrac{e^{-\beta(E_i - \mu N_i)}}{Z_{NVT}}$
Isothermal-Isobaric	N,P,T	$\sum_i e^{\beta PV_i} Z_{NVT}$	$\dfrac{e^{-\beta(E_i + PV_i)}}{Z_{NVT}}$

Each ensemble is associated with a characteristic thermodynamic function. The entropy (S) can be obtained directly from the microcanonical partition function as follows

$$S = k\ln Z_{NVE} \tag{2.7}$$

The Helmholtz function (A) is associated with the canonical ensemble

$$A = -kT\ln Z_{NVT} \tag{2.8}$$

The isothermal-isobaric ensemble yields the definition for the Gibbs function (G)

$$G = -kT\ln Z_{NPT} \tag{2.9}$$

The thermodynamic function for the grand canonical ensemble is pressure (P)

$$P = -kT\ln Z_{\mu VT} \tag{2.10}$$

1.2. Numerical Implementation about Thermodynamic Ensembles

The main object of molecular simulations is to predict atom trajectory in the materials include the atomic position, velocity and acceleration. As for thermodynamic ensemble *NVE* with constant particle number, the volume of system and the total energy of system, the atomistic positions can be acquired by integrated the following equations:

$$r_i(t + \Delta t) = r_i(t) + v_i(t)\Delta t + \frac{F_i(t)}{2m}\Delta t^2 \tag{2.11}$$

$$v_i(t + \Delta t) = v_i(t) + \frac{(F_i(t) + F_i(t + \Delta t))}{2m}\Delta t \tag{2.12}$$

here $F(t + \Delta t)$, $v_i(t + \Delta t)$ and $r_i(t + \Delta t)$ are force, velocity and position of the *i*th atom at the moment $t + \Delta t$ while $F_i(t)$, $v_i(t)$, v_i^n and $r_i(t)$ are force, velocity and position of the *i*th atom at the moment t, Δt is timestep, *m* is mass of atom. Other thermodynamic ensembles such as *NVT* or *NPT* ensembles can be realized by modifying the equations of motion in an appropriate way. Take the example of simple algorithm on enabling *NVT* ensemble, the approach of modifying the equations of motion to obtain a specific thermodynamic ensemble is illustrated. The basic idea of this approach is to change the velocities of the atoms so the temperature can be adjusted to the desired value which is realized as following

$$\lambda = \sqrt{1 + \frac{\Delta t}{\tau}\left(\frac{T}{T_{sec-1}}\right)} \tag{2.13}$$

here Δt is the time step and τ is a parameter which can describes the strength of the coupling to the heat bath. The velocities can be rescaled according to this parameter and given by

$$v_{new,i} = \lambda v_i \tag{2.14}$$

for every atom *i*. There are many other methods to enable the *NVT* ensemble such as *Langevin* dynamics and the *Nose-Hoover* heat bath. The *Parrinello-Rahman* method is suitable for enabling *NPT* ensembles, in which it can adjust both of the temperatures and the pressure to the desired value of the simulated system. During the integration of the equations of motion, molecular dynamics naturally samples the microscopic configuration and provides a collection of snapshots that after averaging correspond to the proper macroscopic state because every microscopic state has a certain probability associated with the corresponding energies. After that, the obtained trajectories can then be used to calculate thermodynamic properties by simply averaging over the sampled configuration.

2. INTERATOMIC POTENTIAL FUNCTION

Whenever a problem is considered at an atomic scale, as in MD simulation of nanometric cutting, there is a need to consider the forces that exist between the molecules (atoms), for it is these forces that decide much of what happens in any phenomenon. Intermolecular forces can be loosely classified into various categories. One may divide them into categories of "opposites"-in other words, attractive or repulsive, short-ranged or long-ranged, strong or weak, and isotropic or directional. These are important distinctions, but they are also ambiguous and confusing. A force that has the same physical origin may be both short-ranged and long-ranged, or it may be attractive in one solvent and repulsive in another. This way of classification can also result in the same force being "counted twice" in any theoretical analysis. To avoid such pitfalls, it is best to classify forces according to their different physical or chemical origin, although even here we shall see that the forces among large particles or extended surfaces lend themselves to different modes of classification from those occurring between two atoms or molecules; this is because of the collective interaction among many molecules, which always includes entropic effects and cannot be easily related to the individual pair potentials.

A potential that is developed for one class of material in all likelihood would not be satisfactory for application to other classes of materials, because the interacting forces are so different. It is therefore necessary to develop a potential for each class of material. Unfortunately, the development of a potential is not a simple, straightforward approach and may take considerable time and expertise. Fortunately, potentials were developed for limited classes of materials and research is in progress for other materials. For example, Morse and Lennard-Jones potentials were applied initially to address cubic metals. Embedded atom potential was developed as an improvement for a wider range of metals. Similarly, Brenner potential and Tersoff potential were developed to address covalently bonded materials, such as silicon, germanium and even diamond. The Born-Meyer potential was developed specifically for some ceramics. It should, however, be noted that these potentials were developed for single-phase, single-crystal materials that have one specific form of bonding. For polycrystalline materials, metallic alloys and for materials which are part ionic and part covalent, the development of new and more complex potentials is required. In this section, some of the empirical potentials are reviewed briefly.

It may be pointed out at the outset that the accuracy of the trajectories of atoms obtained from the MD simulation is highly affected by the choice of the appropriate potential energy (PE) function. Hence, selection of an appropriate PE function is a prerequisite. The total energy of a system is the sum of kinetic energy and potential energy. The kinetic energy is simple to compute while the computation of PE is rather complex as it depends on the position of all interacting atoms. The potential energy plays a central role in MD simulation. Firstly, the force acting on each atom is proportional to the first derivative of the PE function. Secondly, the total energy must be carefully monitored in MD simulation. There are two approaches for the determination of interatomic potentials. The first approach is an *ab initio* method and the second approach uses essentially empirical potentials. In the *ab initio* method, the parameters of the PE function can, in theory, be determined by solving Schrodinger's wave equation. However, in practice, it is difficult to solve by this method except for very simple systems. It may be pointed out that the term "empirical" in empirical potential may be somewhat misleading or oversimplifying as it may not be strictly as empirical as the term suggests. In fact, these potentials often present a more realistic view of the atomic interactions than potentials derived exclusively at great efforts from purely theoretical considerations which are themselves often approximate in nature. The empirical potentials are based on simple mathematical expressions for the pairwise interaction between two atoms or ions, which may or may not be justified from theory, and which will contain one or more parameters adjusted to the experimental data. The validity of the function as well as the stability of the crystal for a given material are checked for various properties including cohesive energy, the Debye temperature, the lattice constant, the compressibility and the elastic constants as well as the equation of state. Consequently, these potentials can be considered as reasonably valid for simple cubic metals. For larger systems, empirical PE functions are used that take into account factors such as covalent bond stretching and bond angle change due to bending, torsion and non-bonded (van der Waals and Coulomb) interactions. This is the second and most common method in which the parameters are determined on the basis of the physical properties of each material. The parameters can be obtained either from experimental studies or using quantum mechanical calculations. The interatomic potential energy is usually assumed to be the sum of *n*-body (empirical) potentials which depend only on the distance between the atoms. These potentials are further classified into two-body, three-body and multi-body potentials according to the unit of atoms on which the potential terms depend. In the following discussion, some empirical potential that are most widely used are given to aid the reader in the selection of appropriate potential for a given application. The attractive force binds the atoms together while the repulsive force prevents them from coalescing. The magnitude of both forces increases as the distance between them decreases, the repulsive force increasing more rapidly than the attractive force. The curvature of the potential energy function is determined mainly by the repulsive force, which therefore dictates the elastic behavior of the solid. The length of the bond is the center-to-center distance of the bonding atoms. Strong bonds pull the atoms closer and so have smaller bond lengths compared with weak bonds. At some definite distance, the attractive and the repulsive forces exactly balance and the net force is zero. This corresponds to stable equilibrium with a minimum potential energy, the magnitude of which is the bond energy. The cohesive properties of the solid, its melting and vaporization behavior are determined by the magnitude of the maximum binding energy, which is governed by the attractive component of the interatomic force. The PE functions generally extend over a modest range of pair separations though at vanishingly small energy levels. By neglecting the interatomic interactions beyond a cut-off point, a significant reduction in the computational time with minor loss in accuracy can be accomplished. Truncation of the potential to a cut-off point also results in a similar truncation in the force curve. The cut-off distance can be of any value of choice but is generally taken as the distance where the value of the potential energy is about 3-5 percent of the equilibrium potential energy.

2.1. Pair Potential Function

Most early molecular dynamics simulations use empirical radically symmetric pair potentials to describe the interactions among atoms. The interaction energy and force between atom and their neighbors is given by the sum of each of the pairwise contributions while no additional cohesive pseudo-potential contributions are involved. The parameters in the expressions are obtained by fitting empirical potential functions to intrinsic materials parameters such as the elastic modulus, crystal structure, cohesive energy, stacking fault energy to bulk properties. The underlying potential functions are usually of arbitrary exponential or higher order polynomial form. The pair potential function usually used to simulate basic structural aspects such as lattice defects and their dynamics. They are particularly preferred when a large number of atoms must be considered because of their simple mathematical structure. Owing to the physical meaning of the empirical terms involved, pair potentials request a good characterization of monoatomic closed-shell gases. The atom-atom pair interaction potential energy only determined by the distance between two atoms and the total energy of the system can be acquired by summing the energy of all atomic bonds over all n atoms in the system which is given by

$$E_{total} = \frac{1}{2} \sum_{i \neq j=1}^{n} \sum_{j=1}^{n} \phi_{ij}(r_{ij}) \tag{2.15}$$

here r_{ij} is the distance between atoms i and j, the factor $1/2$ is to account for the double counting of atomic bonds. The term $\phi_{ij}(r_{ij})$ is the potential energy of a bond formed between two atoms and expressed as a function of its distance r_{ij}.

Presently, there are two types of pair potentials. The first group is referred to as classical pair potentials which describe the total energy of the system and no cohesive term is involved. The second group is referred to as isotropic weak pseudo-potentials which describe the energy change of the system associated with structural changes and include a further cohesive term. Classical pair potentials fully determine the total energy of a system without considering any further cohesive terms that arise from the interaction with atoms far away from the particle considered. The atoms are regarded as mass points which have a central interaction with their nearest neighbors. The interactions of any pair of atoms depends only on their spacing which means that such potentials are principally radially symmetric and independent of the angular position of other atoms in the vicinity.

2.1.1. Morse Potential

Morse potential is a commonly used empirical pairwise PE function for bonded interactions. This potential produces repulsive force in the short range, attractive force in the medium range and decays smoothly to zero in the long range. It uses a form of potential containing two exponential terms instead of power law dependence considered by other researchers. The Morse potential energy function is given by

$$\phi(r_{ij}) = D\left(e^{-2\alpha(r_{ij}-r_c)} - 2e^{-\alpha(r_{ij}-r_c)}\right) \tag{2.16}$$

here r_{ij} and r_c are the instantaneous and equilibrium distances between atoms i and j respectively, D and α are constants determined on the basis of the physical properties of the material. In practical, r_c, α and D are obtained from the closest spacing between the atoms (equilibrium lattice spacing), the Debye temperature and the sublimation energy respectively. The validity of the function as well as the stability of the crystal for a given material are checked for various properties including cohesive energy, the lattice constant, the compressibility and the elastic constants as well as the equation of state and stability of the crystal.

2.1.2. Lennard-Jones Potential

Another simple pair potential for metallic materials corresponding to the hard-sphere model (similar to the Morse potential) that is commonly used is the Lennard-Jones potential. Like the Morse potential, this potential attempts to account both for short-range, repulsive overlap forces and for longer-range, attractive dispersion forces. The resulting Lennard-Jones potential is given by

$$\phi(r_{ij}) = 4\varepsilon\left[\left(\frac{\sigma}{r_{ij}}\right)^6 - \left(\frac{\sigma}{r_{ij}}\right)^{12}\right] \tag{2.17}$$

here σ is determined by the physical properties of the material. By convention, repulsive forces are considered positive while attractive forces negative.

2.1.3. Born-Meyer Potential

A number of potentials were developed to consider only the repulsive interaction for small separation distance. The purpose for the development of such a potential is twofold. Firstly, it can be used in cases where the attractive part plays only a minor or insignificant role. Secondly, the long-range attractive component can be represented by another appropriate function. The Born-Meyer potential was introduced to represent the closed-shell repulsion between ionic crystals and is given by the function

$$\phi(r_{ij}) = Ae^{-2\alpha(r_{ij}-r_0)} \tag{2.18}$$

here A and r_0 are constants determined by the physical properties of the material. The repulsive force decays smoothly as the distance r_{ij} increases. This potential is developed specifically to address ceramic materials.

2.2. Multibody Potential Function

The electrostatic and van der Waals energy reviewed above are calculated between pairs of interaction sites. The total interaction energy is thus determined by adding together the interactions between all pairs of sites in the system. However, the interaction between two molecules (atoms) can be affected by the presence of a third, fourth or more molecules. For example, the interaction energy between three molecules A, B and C is not in general given by the sum of the pairwise interaction energies, namely: $\phi(A,B,C) \neq \phi(A,B) + \phi(A,C) + \phi(B,C)$. Three-body effects can significantly affect the dispersion segment of interaction. For example, it is believed that three-body interactions account for approximately 10% of the lattice energy of crystalline argon. For precise work, interactions involving more than three atoms may have to be taken into account, but they are usually small enough to be ignored. Typical potential function includes both two- and three-body interactions should be given as the following expression

$$\phi(r_{ij}) = \sum_{i=1}^{n}\sum_{j=i+1}^{n}\phi^{(2)}(r_{ij}) + \sum_{i=1}^{n}\sum_{j=i+1}^{n}\sum_{k=j+1}^{n}\phi^{(3)}(r_{ij},r_{ik},r_{jk}) \tag{2.19}$$

The three-body contribution may also be modeled using a term of the form $\phi^{(3)}(r_{ij}) = K_{i,j,k}\left(e^{-\alpha r_{ij}}e^{-\beta r_{ik}}e^{-\gamma r_{jk}}\right)$ where K, α, β, γ are constants describing the interaction between atom i, j and k. Such a functional form has been used in simulation of ion-water systems, where polarization alone does not exactly model configurations when there are two water molecules close to an ion. The three-body exchange repulsion term is thus only calculated for ion-water-water trimmers when the species are close together.

Furthermore, the computational effort is significantly increased if three-body terms are included in the model. The number of bond, angle and torsional terms increase approximately with the number of atoms (n) in the system, but the number of non-bonded interactions increase with n^2. There are $n(n-1)/2$ distinct pairs of interactions to evaluate for a pairwise potential. If three-body effects are included then there are $n(n-1)(n-2)/6$ unique three-body interactions. In general, there are approximately $n/3$ times more three-body terms than two-body terms and so it is clear why it is often considered preferable to avoid calculating the three-body interactions.

2.2.1. Embedded-Atom Potential

Multi-body potentials, such as the embedded-atom (or pair functional) potentials for cubic materials have recently been developed to describe the metallic bonding character more accurately than is possible with two-body potentials. The embedded-atom potential is an extension of the two-body potential for metals and is expected to account realistically for the effect of free electron gas that surrounds each atom. It is considered to be a more realistic PE function that can correctly model the change in properties of metals near a free surface. The total energy of the system is expressed by the embedded-atom potential as follows

$$\phi(r_{ij}) = \frac{1}{2}\sum_{j\neq i}\phi_{\alpha\beta}(r_{ij}) + F_{\alpha}\left(\sum_{j\neq i}\rho_{\alpha}(r_{ij})\right) \tag{2.20}$$

here F is the embedding energy which is a function of the atomic electron density, ϕ is a pair potential interaction, and α and β are the element types of atoms i and j. The multi-body nature of the EAM potential is a result of the embedding energy term. Both summations in the formula are over all neighbors j of atom i within the cutoff distance. With equation (2.20), the interaction force between atoms can be obtained by calculating the negative gradient of $\phi(r_{ij})$. The first term characterizes the conventional pair potential and the second term is the energy to embed atom i in a background of an electron cloud. Thus in this potential, the pair potential is augmented by an additional pairwise sum. As a result, this method can incorporate coordination dependence of the interactions. It is also reported to be computationally efficient to allow for substrate relaxation where needed. A serious limitation is the range of metallic systems for which this method is most accurate. The method works best for metals with completely empty or filled d-bands. They are less reliable for transition metals, near the center of the transition series which have partially filled d-bands. Since these methods have spherically symmetric interactions, they are also not accurate for the central transition metals. However, a modified embedded-atom model (MEAM) potential considers non-spherical or polarized bonding, thus overcoming this limitation. In fact, this model only fit for a number of fcc, bcc and hcp metals. According to this model, there are 11 parameters altogether in the MEAM potential, of which three can be fixed at nominal values. Each of the remaining eight parameters is directly linked to a physical quantity, namely, the sublimation energy, the lattice constant, the bulk modulus, two shear constants, two structural energy differences and the vacancy formation energy. Baskes [6] determined and tabulated the values of the various parameters used in MEAM potential for several cubic metals (bcc, fcc and hcp) which can be used directly for MD simulations.

2.2.2. Tersoff Potential Function

$$E = \sum_i E_i = \frac{1}{2} \sum_{i \neq j} u_{ij} \tag{2.21a}$$

$$u_{ij} = f_C(r_{ij})[f_R(r_{ij}) + b_{ij} f_A(r_{ij})]$$

here E is the total potential energy of the system, f_R and f_A are the repulsive and attractive pair potentials, f_C is a cutoff function given by:

$$f_R(r_{ij})=A_{ij}exp(-\lambda_{ij}r_{ij})\tag{2.21b}$$

$$f_A(r_{ij})=-B_{ij}exp(-\mu_{ij}r_{ij})$$

$$f_C(r_{ij})=\begin{cases} 1, r_{ij}<R \\ \dfrac{1}{2}+\dfrac{1}{2}\cos\left[\pi\dfrac{(r_{ij}-R)}{(S-R)}\right], R<r_{ij}<S \\ 0, r_{ij}>S \end{cases}$$

here r_{ij} is the distance between atom i and j. λ_{ij}, μ_{ij} and R are the potential parameters. The main idea of this potential is that the strength of each bond depends upon the local environment and is lowered when the number of neighbors is high. This dependence is expressed by the parameter b_{ij} which can diminish the attractive force relative to the repulsive force.

$$b_{ij}=\chi_{ij}(1+\beta_i^{n_i}\varsigma_{ij}^{n_i})^{-1/2n}\tag{2.21c}$$

$$\varsigma_{ij}=\sum_{k\neq i,j}f_C(r_{ik})\omega_{ik}g(\theta_{ijk})$$

$$g(\theta_{ijk})=1+\frac{c_i^2}{d_i^2}-\frac{c_i^2}{[d_i^2+(h_i-\cos(\theta_{ijk}))^2]}$$

here ς_{ij} defines the effective coordination number of atom taking into account the relative distance between two neighbors (r_{ij}, r_{ik}) and the bond angle (θ_{ijk}). The function $g(\theta)$ has a minimum at $h=\cos(\theta)$. The parameter d determines how sharp the dependence on the angle is and c expresses the strength of angular effect. The parameters R and S are not optimized but chosen so as to include first neighbors only for several selected high symmetry structures such as silicon, diamond and so on. The parameter χ_{ij} strengthens or weakens heteropolar bonds in multi-component systems. The potential was first calibrated for silicon and then for carbon. The parameters were chosen to fit theoretical and experimental data, such as cohesive energy, lattice constants and bulk modulus obtained from realistic and hypothetical configurations.

3. EQUATIONS OF MOTION

In this section, numerical simulation is adopted to solve the classical equations of motion for a system composed of N molecules interacting through potential function. In reality, numerical methods to solve sets of differential equations can be found in almost any general textbook on applied mathematics and other specialized books. All of these methods are based on finite differences and solve the equations step by step in time while the step size is taken to be constant. The computation of the force is extremely laborious compared to any manipulation involved in updating the variables to take one step forward in time. This means that any method involves more than one force evaluation per step cannot be a reasonable choice. Another requirement is that an algorithm should behave well for the type of force that one encounters in molecular dynamics. Normally molecular dynamics produces a trajectory in the valley regions of a potential field that has a very mountainous character. In the valley regions the second derivatives of the potential with respect to most coordinates are positive. This means that the second spatial derivatives of the potential have a systematic sign and the algorithm should at least incorporate the proper treatment of the first derivate of the forces to avoid accumulating errors and instability of the solution. This means that the order of the algorithm which defined as the highest order of the time step h included in the equation for the solution of the coordinates should be at least 3. Higher orders correspond to higher than first derivatives of the forces. It is expected that these have a more erratic and nonsystematic behavior. In complex systems that also involve cut-off errors in the forces, it is generally not very advantageous to go beyond third or fourth order.

The general idea about atomic dynamic trajectory in the molecular simulation is as follows. Given the molecular positions, velocities and other dynamic information at time t, we attempt to obtain the positions, velocities *etc.* at later time $t + \delta t$ to a sufficient degree of accuracy. The equations are solved on a step by step basis, the choice of time internal δt will depend somewhat on the method of solution, but δt will be significantly smaller than the typical time take for a molecule to travel its own length. Many different algorithms fall into the general finite difference pattern and several will be reviewed in the following section.

Deriving and solving differential equations are thus among the most common tasks in modeling material systems. Differential equations are equations involving one or more scalar or tensor dependent variables, independent variables, unknown functions of these variables, and their corresponding derivatives. Equations which involve unknown functions that depend on only one independent variable are referred to as ordinary differential equations. If the equations involve unknown functions that depend on more than one independent variable they are referred to as partial differential equations. The "order" of a differential equation is the highest order of any of the derivatives of the unknown functions in the equation. Equations involving the second derivatives are referred to as second-order differential equations. Second- and higher-order differential equations such as

$$\frac{d^2 u(t)}{dt^2} = f(u,t) \tag{2.22}$$

can be transformed into a couple set of low-order equations by substitution:

$$\frac{dv(t)}{dt} = f(u,t) \tag{2.23}$$

$$v = \frac{du(t)}{dt}$$

here u is the state variable which is a function of the independent time variable t, v is the first time derivative of u, and f a function of u and v, respectively. For instance, the frequently occurring problem of (one-dimensional) motion of a particle or dislocation segment of effective mass m under a force field $f(x,t)$ in the x-direction is described by the second-order differential equation

$$m \frac{d^2 x(t)}{dt^2} = f(x,t) \tag{2.24}$$

Differential equations which contain only linear functions of the independent variables are called "linear differential equations". For these equations the superposition principle applies. That means linear combinations of solutions which satisfy the boundary conditions are also solutions to the differential equation satisfying the same boundary conditions. Differential equations which involve nonlinear functions of the independent variables are denoted as "nonlinear differential equations". For such equations the superposition principle does not apply. Most problems in molecular modeling lead in their mathematical formulation to "partial differential equations" which involve both space and time as independent variables. Usually, one is interested in particular solutions of partial differential equations which are defined within a certain range of the independent variables and which are in accord with certain initial-value and boundary-value conditions. In this context it is important to emphasize that a problem which is in the form of a differential equation and boundary conditions must be well posed. That means only particular initial and boundary conditions transform a partial differential equation into a solvable problem.

The solution of partial differential equations by use of analytical methods is only possible in a limited number of cases. Thus, one usually has to resort to numerical methods. In the following sections a number of techniques are presented that allow one to obtain approximate numerical solutions to initial- and boundary-value problems. Numerical methods to solve complicated initial-value and boundary-value problems have in common the discretization of the independent variables (typically time and space) and the transformation of the continuous derivative into its discontinuous counterpart, *i.e.*, its finite difference quotient. Using these discretization steps amounts to recasting the continuous problem expressed by differential equations with an infinite number of unknowns, *i.e.*, function values, into a discrete algebraic one with a finite number of unknown parameters which can be calculated in an approximate fashion.

Numerical methods to solve differential equations which are essentially defined through initial rather than boundary values, *i.e.* which are concerned with time derivatives, are often referred to as finite difference techniques. Most of the finite difference simulations addressed in this eBook are discrete not only in time but also in space. Finite difference methods approximate the derivatives that appear in differential equations by a transition to their finite difference counterparts. This applies for the time and the space derivatives. Finite difference methods do not use polynomial expressions to approximate functions. Since any simulation must balance optimum calculation speed and numerical precision, it is not reasonable generally to favor one out of the many possible finite difference solution techniques for applications in simulation of micro & nanomachining. For instance, parabolic large scale bulk diffusion or heat transport problems can be solved by using a simple central difference Euler method, while the solution of the equations of particle motion in molecular dynamics is usually achieved by using the Verlet or the Gear predictor-corrector method. In most cases, it is useful to select a discretization method with respect to the properties of the underlying differential equations, particularly to the highest occurring order of derivative. A second group of numerical means of solving differential equations comprises the various finite element methods. These methods are designed to solve numerically both complex boundary-value and initial-value problems. They have in common the spatial discretization of the area under consideration into a number of finite elements, the temporal discretization in cases where time-dependent problems are encountered, and the approximation of the true spatial solutions in the elements by polynomial trial functions.

Although the finite difference and the finite element techniques can handle space and time derivatives, the latter approach is more sophisticated in that it uses trial functions and a minimization routine. Thus, the finite difference techniques can be regarded as a subset of the various more general finite element approximations. Many finite difference and particularly most finite element methods are sometimes intuitively associated with the solution of large-scale problems. Although this association is often true for finite element methods which prevail at solving meso- and macroscale boundary-value problems in materials science, it must be underlined that such general associations are inadequate. Finite difference and finite element methods represent mathematical approximation techniques. They are generally not intrinsically calibrated to any physical length or time scale. Scaling parameters are introduced by the physics of the problem addressed but not by the numerical scheme employed to solve a differential equation. Owing to this fact, the fundamentals of the various finite difference and finite element techniques are not presented together with the macro-, meso-, micro-, or nanoscale simulation methods discussed in the ensuing chapters, but treated separately in this eBook.

3.1. Discretization of Time

The finite element and the finite difference methods are related in that they both allow one to discretize partial differential equations and solve them under appropriate initial- and boundary-value conditions. However, since the finite difference methods do not require the use of polynomial trial functions and a minimization procedure, they are presented here at the beginning in order to provide an introductory overview of numerical approaches to solve differential equations. Since finite difference methods were originally designed to solve initial-value problems, the present section concentrates on the numerical approximate on of functions that are described by differential equations which contain time derivatives. When numerically solving initial-value problems one encounters two fundamental problems that the computer cannot handle: first, the mapping of the independent variable time, which is continuous rather than discrete, and second, the evaluation of derivatives which are defined through a limit process. Consequently, each finite difference technique is based on two main numerical approximations, namely, the discretization of time from a continuous measure into tiny intervals of length $h = \Delta t$, and the replacement of the differential equation by the corresponding difference equation. The first derivative du/dt of a function $u(t)$ is at t_0 defined through the limit of a difference quotient:

$$\left. \frac{du}{dt} \right|_{t=t_0} = \lim_{h \to 0} \frac{u(t_0 + h) - u(t_0)}{h} \tag{2.25}$$

This continuous derivative can be transformed into the finite difference expression as follows:

$$\frac{\Delta u}{h} = \lim_{\Delta t \to h \neq 0} \frac{u(t_0 + h) - u(t_0)}{h} \tag{2.26}$$

3.2. Numerical Errors of Finite Difference Methods

The simplifications generally inherent in any finite difference method, *viz.*, the discretization of time and the replacement of the differential quotient by its difference quotient, can entail various types of numerical errors. The "truncation error" is the amount by which the exact solution of the continuous differential equation fails to satisfy the approximate equation. It is calculated as the norm of the difference between the solution to the continuous differential equation and the solution to the discretized difference equation divided by the time interval used in the numerical algorithm.

It is clear that the deviation between the analytical solution obtained for the continuous differential equation and that obtained numerically from the approximate equation should degrade by gradually refining the time step of the calculation. This property is referred to as the "consistency" of an approximate method. A finite difference approximation is called consistent with the differential equation if the truncation error goes to zero as the time interval goes to zero. If all approximate solutions to a given differential equation are consistent, the numerical scheme is called a "convergent" method. In this context it should be added that in the case of space-discretized finite difference simulations, such as are encountered in the fields of dislocation dynamics, diffusional phase transformations, and recrystallization, there can be considerable residual errors due to the discretization of space, irrespective of the decrease in the time intervals. Because of the limited precision of computers another numerically introduced error must be considered, namely, the "round-off error". This error is not identical to the truncation error. While the latter stems from the limitation in transforming the difference quotient into a differential quotient, the former is due to the discretization of time. In contrast to truncation errors which are generally reduced with a decreasing time interval, round-off errors can even increase with decreasing time intervals. This becomes clear in a simple example where the time steps are physically given by real numbers in units of seconds, but numerically for some reason these are reduced to integers. For very small time steps it is clear that such round-off errors can entail substantial accumulated deviations from the analytical result. In order to keep this error as small as possible, finite difference computer codes should generally work with double precision. A further important criterion in evaluating the quality of a finite difference scheme is its stability. The investigation of the stability of the solution quantifies whether the errors discussed above will accumulate and amplify without limit during the computation or not.

3.3. Several Classical Finite Differential Methods

3.3.1. Euler Method

As a starting point for explaining the various finite difference schemes it is useful to recall a simple initial value problem:

$$\frac{du}{dt} = f(u,t) \tag{2.27}$$

here the state variable u depends only on the independent time variable t. When the initial condition is given by $u(t_0) = u_0$, the above equation can be written in integral form,

$$u(t) = u_0 + \int_{t_0}^{t_n} f(s,u(s))ds \tag{2.28}$$

The general method to solve the above integral consists in dividing the interval $[t_0, t_n]$ into a larger number, n, of equally spaced subintervals of length $h=(t_n-t_0)/n$. Partitioning the time in the above integral into discrete intervals which leads to a discretized equivalent of equation:

$$u_t = u_0 + \sum_{j=0}^{n-1} \int_{t_j}^{t_{j+1}} f(s,u(s))ds \tag{2.29}$$

The simplest method to approximate the time derivative du/dt at the time t_i is to identify the local tangent at point t_i with the slope calculated from the two subsequent values of the state variables $u_i = u(t_i)$ and $u_{i+1} = u(t_{i+1})$, and the time interval $h = t_{i+1} - t_i$.

$$\frac{du(t_i)}{dt} \approx \frac{u_{i+1} - u_i}{h} \tag{2.30}$$

Quantifying the truncation error shows that it is linearly proportional to the chosen time interval h. This means that equation (2.30) can be rewritten

$$\frac{du(t_i)}{dt} = \frac{u_{i+1} - u_i}{h} + O(h) \tag{2.31}$$

Combining equation (2.31) with equation (2.27) rewritten for the time t_i

$$\frac{du(t_i)}{dt} = f(u_i, t_i) \tag{2.32}$$

Leads to

$$u_{i+1} = u_i + hf(u_i, t_i) \tag{2.33}$$

This method is referred to as "forward" or "explicit" *Euler* method since the value u_{i+1} is given directly in terms of some previously computed value of the state variable u_i. The "backward" or "implicit" *Euler* method differs from the explicit method in the evaluation of f in (t_{i+1}, u_{i+1}) rather than in (t_i, u_i). The value of the state variable u after $(i+1)$ steps then amounts to

$$u_{i+1} = u_i + hf(u_{i+1}, t_{i+1}) \tag{2.34}$$

The implicit Euler method generally leads to a similar truncation error as the explicit approach, *i.e.*

$$\frac{du(t_i)}{dt} = \frac{u_i - u_{i-1}}{h} + O(h) \tag{2.35}$$

Although the implicit *Euler* algorithm looks at first very similar to the explicit method, it is computationally less efficient. This is evident from equation (2.34), where the unknown u_{i+1} appears on both sides of the equation, *i.e.*, the expression must be transformed before its solution. This is not the case for the explicit *Euler* method, equation (2.33), which can be solved directly since the value of u_i is already known. However, the implicit *Euler* method is more stable than the explicit one and is thus preferred in certain cases irrespective of its lower efficiency. The various *Euler* methods can also be derived by using the Taylor formula which allows one to express the change in the state variable $u(t_0 + h)-u(t_0)$ in terms of a series expansion

$$u(t_0 + h) - u(t_0) = \sum_{j=1}^{n-1} \frac{h^j}{j!} \frac{d^j u(t_0)}{dt^j} + \varsigma_n \tag{2.36}$$

where ς_n is the residual error of the expansion in the case of $n < \infty$. Conducting the Taylor expansion for three subsequent equidistant points $t_{i-1} = t_i\text{-}h$, t_i and $t_{i+1} = t_i + h$ leads to three equations:

$$u_{i-1} - u_i = -h\frac{du_i}{dt} + \frac{h^2}{2!}\frac{d^2u_i}{dt^2} - \frac{h^3}{3!}\frac{d^3u_i}{dt^3} + \frac{h^4}{4!}\frac{d^4u_i}{dt^4} - \cdots$$

$$u_i - u_i = 0 \tag{2.37}$$

$$u_{i+1} - u_i = h\frac{du_i}{dt} + \frac{h^2}{2!}\frac{d^2u_i}{dt^2} + \frac{h^3}{3!}\frac{d^3u_i}{dt^3} + \frac{h^4}{4!}\frac{d^4u_i}{dt^4} + \cdots$$

Dividing the last of these equations by h leads to an expression similar to that obtained for the forward *Euler* method:

$$\frac{du_i}{dt} = \frac{u_{i+1} - u_i}{h} - \frac{h}{2!}\frac{d^2u_i}{dt^2} - \frac{h^2}{3!}\frac{d^3u_i}{dt^3} - \frac{h^3}{4!}\frac{d^4u_i}{dt^4} - \cdots \tag{2.38}$$

The truncation error in this expansion is characterized by the lowest-order term in h. Equivalent Taylor-type derivations can be made for the backward and the central difference quotients.

3.3.2. Leap-Frog Method

A general reduction of the truncation error as compared with the first-order explicit and implicit Euler methods can be achieved by using the more symmetric second-order "central difference" or "leap-frog" method:

$$\frac{du_i(t)}{dt} = \frac{u_{i+1} - u_{i-1}}{2h} + O(h^2) \tag{2.39}$$

In this method the value of variable u at time t_{i+1} is calculated as following:

$$u_{i+1} = u_{i-1} + 2hf(u_i, t_i) \tag{2.40}$$

Depending on the required accuracy high-order central difference algorithms can also be formulated.

3.3.3. Predictor-Corrector Method

The predictor-corrector method is a finite difference method that in its simplest form includes a forward *Euler* method which is referred to as the "predictor", and a subsequent correction of the predicted result which is named the "corrector". The procedure can be used in an iterative manner. Together with the Verlet algorithm, it represents the most common method of integrating the equations of motion in molecular dynamics. The predictor-corrector method proceeds by first estimating the value of the state variable u at time t_{i+1} using an explicit *Euler* step.

$$u_{i+1}^{\alpha} = u_i + hf(u_i, t_i) \tag{2.41}$$

This step is referred to as the predictor, since it provides a first prediction of the value of u at t_{i+1} which is here denoted by u_{i+1}^{α}. In a second step this value is modified by using the implicit algorithm

$$u_{i+1} = u_i + hf(u_{i+1}^{\alpha}, t_{i+1}) \tag{2.42}$$

This step is named the corrector, since it changes the initial explicit prediction of u_{i+1}. Using the second step, equation (2.41), more than once turns the predictor-corrector method into an iterative technique. This process, which does not affect the order of the finite difference method, is called correcting to convergence:

$$u_{i+1}^{\alpha} = u_i + hf(u_i, t_i)$$

$$n \text{ times} \begin{cases} u_{i+1} = u_i + hf(u_{i+1}^{\alpha}, t_{i+1}) \\ u_{i+1}^{\alpha} = u_{i+1} \end{cases} \tag{2.43}$$

Iterative predictor-corrector methods prevail as finite difference algorithms in molecular dynamics, fluid dynamics, and simulations of diffusion.

3.3.4. Crank-Nicholson Method

The Crank-Nicholson method is a second-order finite difference method. It consists in averaging the values of f in t_i and t_{i+1}. The value of the state variable u at time t_{i+1} amounts to

$$u_{i+1} = u_i + \frac{h}{2} \left[f(u_i, t_i) + f(u_{i+1}, t_{i+1}) \right] \tag{2.44}$$

This method corresponds to proceeding for half the time interval along the derivative determined at time t_i and for the remaining part along the derivative at t_{i+1}. Mathematically, this technique is identical to the trapezoidal rule. The truncation error shows a decay which is proportional to h^2. The *Crank-Nicholson* method can also be transformed into a predictor-corrector formulation where the first evaluation of u_{i+1}^α is achieved by a forward *Euler* method and the second one by using the averaging *Crank-Nicholson* step.

$$u_{i+1}^\alpha = u_i + hf(u_i, t_i)$$

$$n \text{ times} \begin{cases} u_{i+1} = u_i + \dfrac{h}{2}[f(u_i + t_i) + f(u_{i+1}^\alpha, t_{i+1})] \\ u_{i+1}^\alpha = u_{i+1} \end{cases} \qquad (2.45)$$

The iterative (n = 1) predictor-corrector combination of the explicit *Euler* method and the *Crank-Nicholson* step can also be referred to as the second-order *Runge-Kutta* method.

3.3.5. Runge-Kutta Methods

The *Crank-Nicholson* method introduced above, which is essentially based on a modified *Euler* algorithm, can be regarded as a special second-order case of the more general *Runge-Kutta* method. In fact, all the previously discussed methods have in common that they express the solution in terms of the derivative which is calculated for different times. Thus, all *Euler*-based methods are often classified as generalized *Runge-Kutta* algorithms. The *Runge-Kutta* formulation of the *Crank-Nicholson* method, equation (2.45), can be written

$$u_{i+1} = u_i + \frac{h}{2}[F_1 + F_2]$$
$$F_1 = f(u_i, t_i) \qquad (2.46)$$
$$F_2 = f(u_i + hF, t_i + h)$$

In their most general form the functions at different stations amount to

$$F_n = f[u_i + h(A_{n1}F_1 + A_{n2}F_2 + \cdots + A_{nn-1}F_{n-1}), t_i + a_n h] \quad n = 1, \cdots, o \qquad (2.47)$$

where a_n and A_{nm} are coefficients and o is the order of the *Runge-Kutta* approximation. Equation (2.47) reveals a certain similarity to the Taylor formula, which allows one to express the change of the dependent variable $u(t_0+h)-u(t_0)$ as a series expansion. While the Taylor expansion is carried out on *one* point, the Runge-Kutta method uses derivatives on *various* points. However, the coefficients a_n and A_{nm} in equation (2.47) can indeed be derived by replacing the functions by Taylor series expansions. Comparing the coefficients of both forms leads to a system of linear equations which can then be solved. Thus, the term with the highest order in the Taylor expansions determines the order of the *Runge-Kutta* algorithm. Following this procedure the coefficients of the fourth-order *Runge-Kutta* formula amount to

$$
\begin{aligned}
u_{i+1} &= u_i + \frac{h}{6}[F_1 + 2F_2 + 2F_3 + F_4] \\
F_1 &= f(u_i, t_i) \\
F_2 &= f(u_i + \frac{h}{2}F_1, t_i + \frac{h}{2}) \\
F_3 &= f(u_i + \frac{h}{2}F_2, t_i + \frac{h}{2}) \\
F_4 &= f(u_i + hF_3, t_i + h)
\end{aligned}
\qquad (2.48)
$$

4. ATOMISTIC MODEL ABOUT MNT

In this section we discuss some fundamental concepts associated with model building and the solution of the MNT using molecular dynamics simulations. Therefore the key parameters of simulation such as the model size, cutting speed and boundary conditions which provide adequate resolution to observe plastic behaviors with reasonable computation time are determined. The idea of using molecular dynamics to precision material removal process results by two related subjects. First, precision machining has steadily decreased in scale from millimeters to angstroms which is closely related to nanotechnology. Secondly, the study of hard and brittle materials such as ceramics and glasses, microscopic phenomenon and structure are important. Brittle materials are difficult to process in a traditional sense due to their tendency to fracture. The fracture can lead to the formation and propagation of cracks which can be readily studied using MD method. MD models will be helpful about the prediction and evaluation of various process variables in an effort to better understand the fundamental processes and phenomena such as materials composition, properties and fabrication parameters. Molecular dynamics is amenable to investigate the micromechanics of hard & brittle materials because of its atomistic resolution.

4.1. Model Size

Model size is important about represent nanometer scale fabrication process such as dominant plastic behaviors and materials removal at atomic scale. Simulation results carried out by the author and other researchers indicate that larger model can provide more detail information of phenomenon while the increasing of model size is constrained by the computation time. Since calculation of interatomic force acting on each atom takes most of computation time, we can expect that total computation time of each simulation depends on the number of workpiece atoms which is proportionate to the product of the length and the height of the workpiece. With the consideration on both the capability of visualizing plastic materials behaviors of interest and the length of computation time, the length and height of the model ought to less than 10 nanometers. Note that the scale of the simulation has been set due to computational considerations and simulation results in the study are specific to nanomachining. Determining the extent to which the results are analogous to micromachining would require a convergence analysis which is out of range of this study.

4.2. Cutting Speed

Along with model size, cutting speed affect computation time because it determines the number of simulation steps when cutting distance and simulation time step is fixed. In order to conduct simulations within reasonable computation time, the cutting speed in the simulation is always increased to speed higher than those typically used in the real machining process. However, cutting speed cannot exceed the speed of sound in the material as the dislocation in the crystal lattice cannot occur beyond the velocity. Furthermore, increasing of cutting speed will have great side effect upon the machining process, 1) an increase in the strain rate; 2) an increase in the cutting temperature. The increase of strain rate will result in an increase of the flow shear stress of the material. Therefore, the overall magnitude of stress in the shear deformation regions, and also cutting forces will increase with cutting speed. Rate effects may play an important role in the process studied at the high rates. Secondly, the high cutting speed will result in marked increase in the workpiece temperature because there is not enough time for the complete heat conduction with surrounding environment. Generally, the main purpose of high cutting speed is to reduce the computation time. Therefore, some method of controlling the temperature of the workpiece has already been utilized in order to desensitize the MD simulation for the temperature effects of the higher cutting speeds. A common method is to place thermostat atoms near the fixed boundary for the cutting heat generated during machining to be properly conducted out the system. However, this approach is not effective enough to stabilize the system because of the high cutting speed and the numerical instability caused by the computational errors. Thus, instead of using thermostat atom, the average kinetic energy of the workpiece is maintained to be consistent with a definite temperature by scaling the velocities of workpiece atoms at each time step which may lead to the unrealistic description of the kinetic energy distribution in the workpiece.

4.3. Thermal Model

Using non-equilibrium MD formulation, the total energy of the system can be partitioned between kinetic energy (KE) which included thermal activation energy and potential energy (PE) or internal strain energy. Temperatures are calculated over some finite number of numerical iterations to produce an average. The initial thermal conditions are based on the starting simulation temperature of room temperature, the thermal energy is input as some combination of KE and PE. Basically, three cases are used for the initial thermal conditions, namely, 1) all KE, 2) all PE, 3) partition the energy. In the author's research, the initial thermal energy is treated as kinetic energy only. Equilibrium simulations are performed to ensure that the system is stable and the energy of the system is conserved. For the harmonic potentials used, small perturbations caused by thermal fluctuations, cause the total energy to tend toward an equipartitioning between the KE and PE components, regardless of the initial thermal conditions. This is due to the fact that for vibrations about the equilibrium positions the potentials are very nearly harmonic in nature.

4.4. Materials Structure

Materials description of MD simulation is different from typical materials used in machining process. Firstly, the workpiece in simulation is in a perfect single crystalline structure which is different from multiphase of multigrain structure of typical materials. Single crystal material is anisotropic and has some preferable slip planes and slip directions. In fact, the single crystalline approximation of the workpiece structure is valid in the case of cutting within a single grain because the engagements of nanoscale machining are significantly smaller than the grain size. Secondly, the yield strength of perfect crystal materials is much greater than those of ordinary materials because shear deformation occurs by the movement of dislocations. This can result in higher cutting forces and high maximum elastic strain which lead to more materials recovers elastically on the clearance face passing under the tool and the engagement of the transition from plough to cutting occurs can be larger than typical materials.

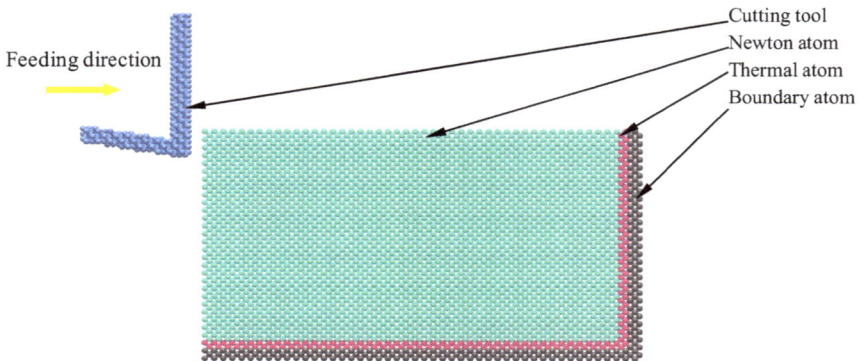

Figure 1: Classical MD simulation model of MNT.

Fig. (**1**) gives the classical MD simulation model of MNT. The workpiece material is divided into three regions, namely, newton atom, thermal atom and boundary atom. There are two kinds of forces acted upon newton atom, one comes from the interaction between workpiece atoms and the other originated by the interaction between cutting tool atom and workpiece atom. The newton region is the key area of the materials removal while there are only internal forces induced by workpiece materials exerted upon thermal atoms. Thermal atoms act as heat boundary which can conduct the cutting heat generated in the machining to the external environment and avoiding the high temperature. Boundary atom act as position boundary and keeps static in the simulation. This region has two functions, one is to keep a definite geometry shape and the other is to prevent the rigid motion.

4.5. Newton Atom

The trajectory of newton atom can be acquired by means of numerical integration:

$$m\frac{d^2\vec{r_i}}{dt^2} = \frac{d(m\vec{v_i})}{dt} = \frac{d\vec{p_i}}{dt} = \vec{F_i} \tag{2.49}$$

here m is the mass of atom, $\vec{r_i}$, $\vec{v_i}$, $\vec{p_i}$ and $\vec{F_i}$ are the position, velocity, momentum and the external force. The external force acted upon atom can be acquired by solving the negative gradient of empirical potential:

$$\vec{F} = -\nabla_i V(\vec{r_1},\vec{r_2},\cdots\vec{r_n}) \tag{2.50}$$

Here V is the potential function, n is the number of the atom, $\vec{r_i} = x_i\,\vec{i} + y_i\,\vec{j} + z_i\,\vec{k}$ is the position vector of the atom i, x_i, y_i and z_i are coordinates of atom i, ∇_i is the Laplace operator.

$$\nabla_i = \frac{\partial}{\partial x_i}\,\vec{i} + \frac{\partial}{\partial y_i}\,\vec{j} + \frac{\partial}{\partial z_i}\,\vec{k} \tag{2.51}$$

There are about *6n* differential equations to be solved about the system composed of *n* atoms. A model includes 2000 atoms requires integration of 12,000 coupled, first-order differential equations of motion. Considering that the potential energy function is a simple pairwise potential form, the total number of terms in such a potential is given by *n(n-1)/2*. Some 2 million pairwise terms need to be calculated each time about 2000 atoms system when a derivative is evaluated. As four such evaluations must be done for every integration step, some 8×10^6

evaluations would be required for every trajectory calculation. The computation time increases rapidly as the number of atoms considered increases.

4.6. Thermal Atom

The plastic deformation energy in the first shear deformation region and the friction induced deformation energy at the interface are converted into heat, which should be continuously transferred into the surrounding environment. In reality, most of the heat is carried away by the chip and the lubricant. It is essential that the effects upon the energy transfer within the solid that would be present for an extended lattice model to be included in the calculations. This would be accomplished by using lattice models containing larger number of atoms. In the molecular dynamics simulation, the temperature of the workpiece will arise higher and higher because of the doing work on workpiece induced by cutting tool. The thermal atom layer is set around the newton atom to mimic the heat transfer and also makes the simulation result more reasonable. Presently, the most efficient method for simulating the removal of the heat generated in machining is the use of velocity reset functions.

The velocity of the thermal atom will be scaled by some definite calibration function at every step:

$$v_i^{new} = (1-W)^{1/2} v_i^{old} + W^{1/2} V(T, \xi) \tag{2.52}$$

here v_i^{old} is the velocity of *ith* atom, v_i^{new} is the calibrated velocity, W is a parameter that controls the strength of the reset, which $W=0$ corresponding to no reset and $W=1$ being a complete reset. $V(T, \xi)$ is a randomly chosen velocity from *Maxwell-Boltzmann* distribution at temperature T. ξ is a random number whose distribution is uniform on the interval $[0,1]$ that controls the random selection. Δt should be approximately 1/5 of the Debye period of the lattice adapted. This procedure simulates the thermostatic effect of the bulk and guarantees that the equilibrium temperature will approach the desired value, which is 293K in these calculations. The initial velocity for each of the atoms in the newton and thermal region is chosen to follow the *Maxwell-Boltzmann* distribution specific to the initial bulk temperature.

4.7. Boundary Atom

The boundary atoms of the workpiece and the atoms of cutting tool are assumed to be unaffected by the materials removal process. Consequently, the position of the boundary atoms with respect to each other will not change during the materials processing. In Fig. (**1**), the tool is composed of only boundary atom which simulates an infinitely hard tool. This is generally the case where a soft workpiece such as copper or aluminum is machined by a hard tool such as diamond.

5. ATOMISTIC SCALE ANALYZING THE MECHANISM OF MNT

5.1. Nanometer Cutting

Fig. (**2**) shows three dimension computation model of nanometer cutting of single crystal silicon. Brittle materials such as silicon, germanium and ceramics have already been widely used in many areas such as computer industry, semiconductor, aerospace and so on. The final machined part of these areas is required to have high form accuracy, better surface quality and little surface (subsurface) damage. But it is difficult to meet this requirement for the physical property of brittle materials such as low fracture strength and so on. Recently, many scholars have verified that brittle materials can be removed in ductile mode under a depth of cut smaller than critical value and then acquire super-smooth machined surface.

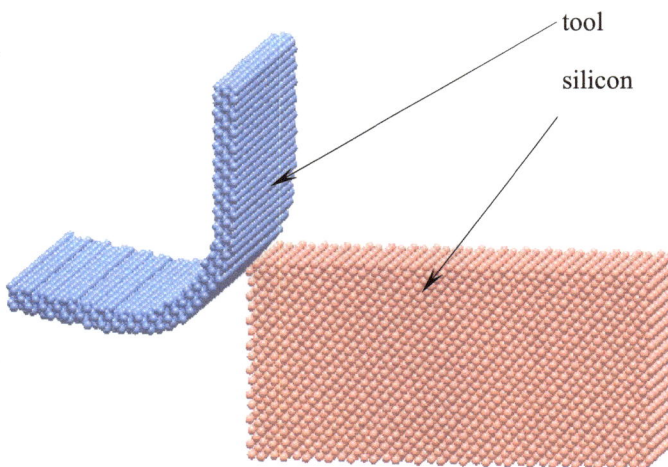

Figure 2: MD Simulation Model.

Although the ductile behavior of silicon under huge hydrostatic pressure has now been widely accepted, the micro-mechanism of phase transformation is still an issue. We investigate the phase transformation of single crystal silicon substrate during nanometer cutting process using rigid tool on the atomic level so as to provide strong basis for constructing macroscopic model. The diamond tool is assumed as ideal rigid body without any deformation. Silicon atoms of workpiece are initially arranged in diamond cubic structure with a constant lattice parameter of 5.43 Å. The cutting speed is 85 m/s, the environment temperature is 293 K, the depth of cut is 10 Å.

For covalent systems, the Tersoff potential is used to depict the interaction between the silicon atoms and between carbon atoms of the cutting tool as follows:

$$\phi_{ij} = f_c(r_{ij})[f_R(r_{ij}) + b_{ij}f_A(r_{ij})]$$

(2.53)

With equation (2.53), the interaction force between silicon atoms can be obtained by calculating the negative gradient of ϕ.

The interaction between the silicon atom and the diamond atom is modeled by the Morse potential as follows:

$$\phi_{ij} = D[e^{-2\alpha(r_{ij}-r_0)} - 2e^{-\alpha(r_{ij}-r_0)}]$$

(2.54)

The parameters such as D, α and r_0 are shown in Table **2**. The interaction force between silicon atom and diamond atom is calculated by the negative gradient of ϕ.

Table 2: Computation Parameters.

Parameters	C-Si
D (ev)	0.435
α (Å^{-1})	4.6487
r_0 (Å)	1.9475

The computation of atom trajectory requires numerical integration of the differential equations of motions from initial state which the cutting tool is approaching the workpiece but has not touched yet to the final state which a layer of material has been removed from the workpiece. There is variety of methods available for performing this numerical integration such as fourth-order Runge-Kutta method, Leap-Frog method, Verlet method, Velocity-Verlet method and so on. The Velocity-Verlet method is a symplectic algorithm which can prevent the energy dissipation and have high computation efficiency, this paper adopted this method as follows:

$$r_i^{n+1} = r_i^n + hv^n + \frac{h^2}{2m} F_i^n \qquad (2.55)$$

$$v_i^{n+1} = v_i^n + h(F_i^n + F_i^{n+1})/2m \qquad (2.56)$$

here r_i^{n+1}, v_i^{n+1} and F_i^{n+1} are position, velocity and force at n+1 step of the ith atom while r_i^n, v_i^n and F_i^n are position, velocity and force at n step of the *i*th atom, *h* is timestep, *m* is mass of atom.

Snapshots of location of the silicon atoms under the conditions of different edge radius during the nanometer cutting process are shown in Fig. (**3 (a)-(g)**). The results show that the nanometer surface generation process consist three steps, namely, compression of work material ahead of tool in the primary deformation region, formation of chip by mechanism similar to extrusion process, subsurface deformation of the material underneath the tool and reconstructing of machined surface. The cutting effect is weakened while the plough and scratching effect is enhanced accompanying the increasing of tool edge radius (Fig. **4**). The transitional stage from initial to stable cutting process is extended and the residual deformation is increased too. The results also indicate that the silicon removal mode is similar to the ductile metal removal mode. But there are also some significant differences about the nature of deformation in metal comparing with silicon because of their different ductility character. Chip formation is induced by generating and propagating of dislocation ahead of the tool as well as into the work material in the case of nanometer cutting single crystal metal while an

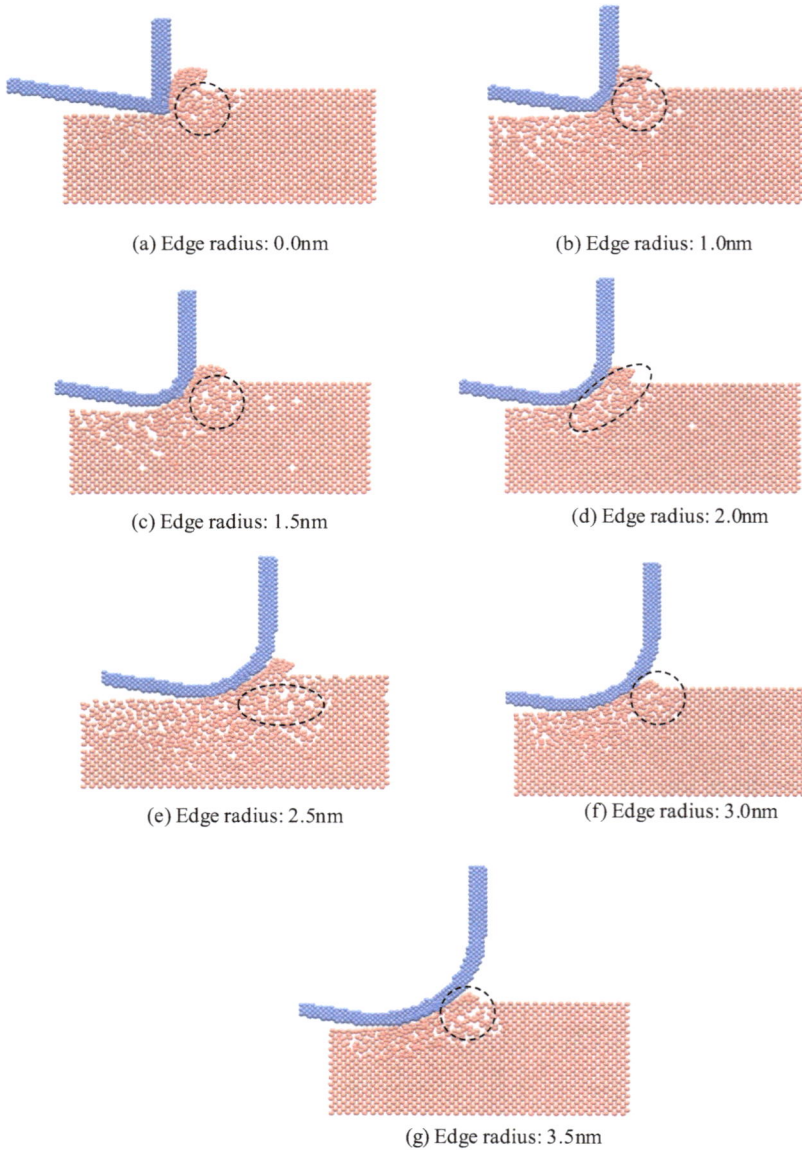

(a) Edge radius: 0.0nm

(b) Edge radius: 1.0nm

(c) Edge radius: 1.5nm

(d) Edge radius: 2.0nm

(e) Edge radius: 2.5nm

(f) Edge radius: 3.0nm

(g) Edge radius: 3.5nm

Figure 3: MD simulation of nanometric cutting with different edge radius.

extrusion-like chip formation process in the case of nanometer cutting single crystal silicon is observed. The single crystal silicon has crystal structure the same as single crystal diamond which belongs to hexahedral crystal system. It has already been widely proved that the break up or wear of this kind of single crystal material has high correlation with cleavage. There may be obvious fracture and cleavage mark if the silicon is removed in brittle mode in the case of nanometer cutting process. However, the ideal silicon atom structure is broken up at tool-workpiece interface, and densification of the material accompanying some new atom structure is observed while no material fracture is happened. The ductile metal deformation character such as propagation of dislocation and slippage is not observed in the workpiece which makes the traditional ductile deformation model unable to explain this material removal mechanism of silicon workpiece. All of the above facts illuminate that brittle material such as silicon can be machined in ductile mode and thus acquiring nanometer level surface in the case of nanometer cutting technology with the depth of cut lies in the range of nanometer or sub-nanometer. There should be some new physical essence dominates the ductile material removal mode for this brittle material and some powerful theoretical toolkit is needed to study its micro-mechanism.

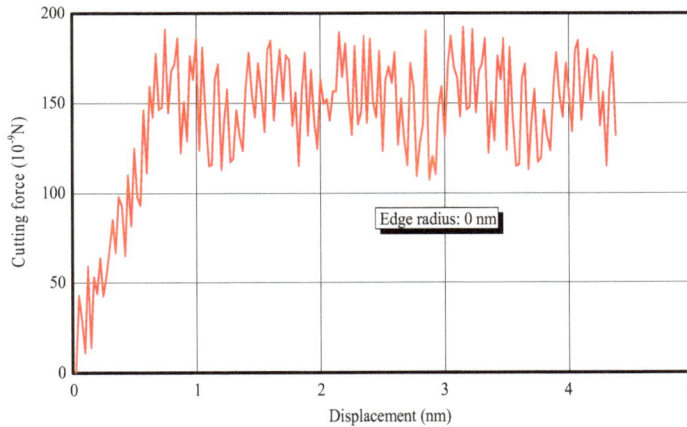

Figure 4: Nanometric cutting force with different edge radius.

Before nanometer cutting, the silicon atom has a bond length of 2.352 Å which is an ideal α silicon structure. In the case of MD simulation nanometer cutting, in order to break up ideal chemical bond between silicon atoms, huge hydrostatic pressure is induced at the local deformation area. This huge pressure exceeds the critical value (10 GPa) of phase transformation thus lead to a large number of new structure which has atomic bond length of 2.43 Å and 2.58 Å. Silicon atoms with three coordinate nearest neighbors are observed due to point defect induced by cutting process and the impending bond in the machined surface. This graph also implies that in the phase transformation process, some silicon atoms are in the flattening tetrahedron of the diamond structure, some silicon atoms complete the phase transformation and are in the β silicon structure while more atoms are in the transition state. The atom with four coordinate neighbors is subjected to small deformation and has a mean potential energy of -4.32 ev. The atom with three coordinate neighbors has a maximum mean potential energy of -2.65 ev. The potential energy of β silicon structure atom is -3.43 ev which far bigger than that of its original diamond structure. This result implies that silicon atom of β structure is unstable than diamond structure. It is also implies the difficulty of phase transformation from diamond structure to β silicon structure (the local hydrostatic pressure exceed 10 GPa) (Fig. **5**).

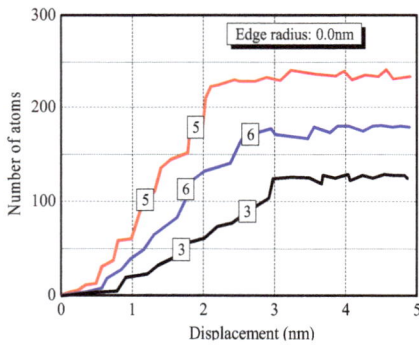

Figure 5: Number of atoms with specified nearest number of neighbors.

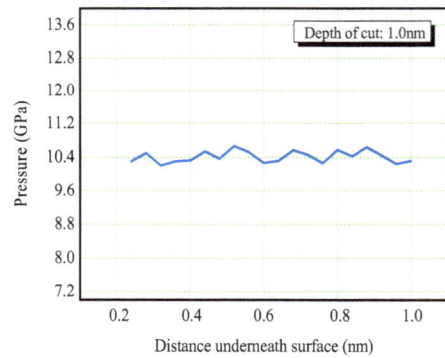

Figure 6: Distribution of pressure beneath the machined surface.

When cutting hard-brittle material, it is the hydrostatic pressure around tool tip determine the deformation property before the material flake off the substrate, namely, the value of hydrostatic pressure dominate the extent of strain and the strain dominate whether plastic deformation or brittle deformation happened in the material. When the depth of cut far less than the critical value, the pressure in the cutting area is uniformly distributed (Fig. (**6**)) induced by the nanometric level tool edge radius. The uniformly distributed pressure in the cutting area can avoid stress concentration and brittle crack thus result in plastic deformation in the material. As the depth of cut far greater than the critical value, the sharp tool edge may result intense stress singularity in the local area and lead to high stress gradient. When the cutting tool moving ahead, the high stress gradient can result in brittle crack and generate concavity in the machined surface.

5.2. Scanning Probe Nanolithography

New developments on integrated-circuit-related technologies have led to rapid progress in the development of microdynamic systems which is known as microelectromechanical systems (MEMS). Mechanical materials for some of the microdynamic systems that have thus far been demonstrated consist of deposited thin films of silicon, silicon nitride, aluminum, polyimide, and tungsten among other materials. Continued research on the mechanical properties of the electrical materials forming microdynamic structures (which previously had exclusively electrical uses), on the scaling of mechanical design, on tribological effects, on coatings, and on the effective uses of computer aids is now under way.

Surface micromachining is an important MEMS microfabrication technique which can be used to create movable microstructures on top of a silicon substrate. The main advantage of surface micromachining is that extremely small sizes can be obtained. In addition, it is relatively easy to integrate the micromachined structures with on-chip electronics for increased functionality. The earliest methods used to build structures from silicon making use of lithography and etch technology that was mastered as a part of solid-state device research was *via* substrate micromachining. The design and fabrication of polysilicon resonant beams together with on-chip MOS circuitry demonstrated the practicality of what has become known as surface micromachining. In contrast to substrate micromachining in which mechanical parts are sculpted from the wafer itself, surface micromachining makes mechanical elements out of materials deposited on the wafer surface.

Nanotribology study is needed to develop fundamental understanding of interfacial phenomena on a small scale and to study interfacial phenomena in surface micro-structure machining. Friction and wear of lightly loaded atomic force microscope (AFM) tip are highly dependent on the surface interactions (few atomic layers). Investigation of nanotribology is valuable about fundamental understanding of interfacial phenomenon, and can provide a bridge between science and engineering. Presently, AFM tip is always used to study single asperity contact based wear/friction with a solid or lubricated surface. AFM become the basic tools used to fabricate the various three-dimensional micro-structures. *In situ* surface characterization of local deformation of materials and thin coatings can also be carried out using a tensile stage inside an AFM. Mechanical properties such as hardness, Young's modulus of elasticity and creep/relaxation behavior can be determined on the micro- to picoscales using a depth-sensing indentation system in an AFM. Localized surface elasticity and viscoelastic mapping can be obtained of near-surface regions with nanoscale lateral resolution. AFM has already become the main important tools of nanofabrication/nanomachining [7]. Single crystal silicon, being typical hard-brittle materials, has always been one of the hard to fabricate materials. There is always microcrack/pit from the depth of cut and the structure dimension lies in the range of nanometer range, materials removal and corresponding plastic deformation must be different from the mechanism at macroscopic scale. This small scale deformation and the tribology behavior are far from forming a convinced explanation. Traditional machining theory cannot uncover the physical essence underline nanofabrication process. Microscopic modeling using molecular dynamics (MD) simulation is carried out to study the materials deformation and tribology behavior at nanometer scale.

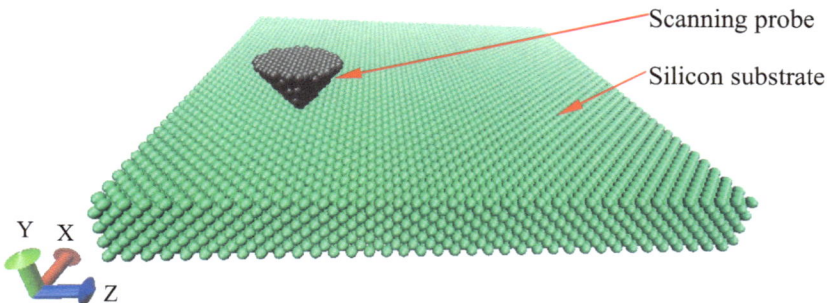

Figure 7: MD Simulation Model.

The atomic configuration of the system is illustrated in Fig. (**7**). Silicon atoms of substrate and diamond atoms of AFM tip are initially arranged in diamond cubic structure with a constant lattice parameter of 5.43 Å and 3.615 Å separately. The dimension of silicon workpiece is $200 \times 50 \times 200 \, \text{Å}^3$ along x, y and z directions. The moving speed is 70m/s, the environment temperature is 293 K, the depth of cut is 8 Å, the timestep is 2.5×10^{-15} s. For covalent systems, the Tersoff potential (equation 2.52) is used to depict the interaction between the silicon atoms, The interaction between silicon atom and diamond atom is depicted by Morse potential (equation 2.53).

Fig. (**9**) gives the MD simulation results of nanolithography of silicon using AFM. The whole process composed of six stages, namely, downward movement (along negative Y axis direction), two pair of parallel linear movement along X and Z axis direction and the upward movement (along positive Y axis direction) which is shown in Fig. (**8**). These three moving directions represent the three series representative crystal orientation, namely, [100], [010] and [001]. There is no brittle crack or dislocation band generated in the surface or at the bottom of shallow groove as shown in the simulation results. The scratch processes described above involve plastic deformation of material in small localized regions where the opposing surfaces or hard particles make contact. It turns out that the onset of plasticity for such small regions can be dramatically different with the plastic yield stress determined by macroscopic measurements. In reality, macro-scale plastic deformation within crystalline materials mainly occurs through the motion of flaws and defects in the crystal structure, which is called dislocations. When the size of the region undergoing plastic flow becomes much smaller than the typical distance between dislocations, however, the contribution of dislocations to the yield stress becomes insignificant; instead, the yield stress is governed by the force needed to slide one plane of atoms over another. Fig. (**10**) shows the cross sectional view of substrate surface after scratching. No obvious brittle crack or dislocation is observed in the materials interior. The volume V_L of materials displaced when the asperity slides a distance L (Fig. (**11**)) is given by:

$$V_L = \frac{1}{2} L h^2 \tan \beta \qquad (2.57)$$

If V_a is the volume of silicon atom, then the number of removed atoms (N) after scratching is:

$$N = V_L / V_a \qquad (2.58)$$

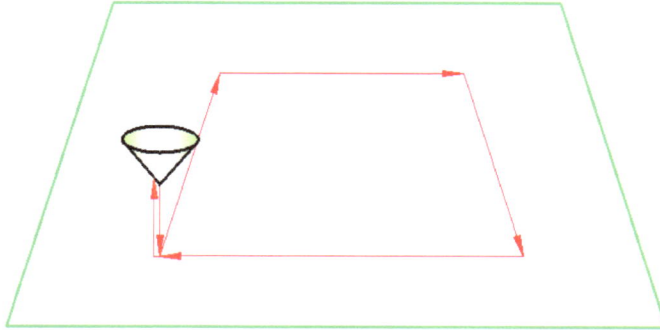

Figure 8: The trajectory of AFM tip.

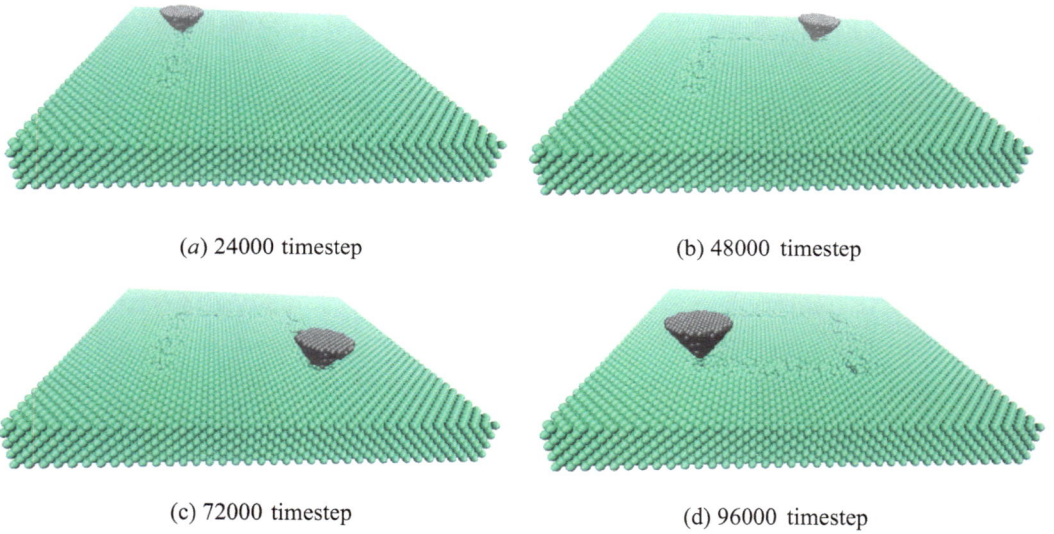

(*a*) 24000 timestep

(b) 48000 timestep

(c) 72000 timestep

(d) 96000 timestep

Figure 9: MD simulation of nanolithography

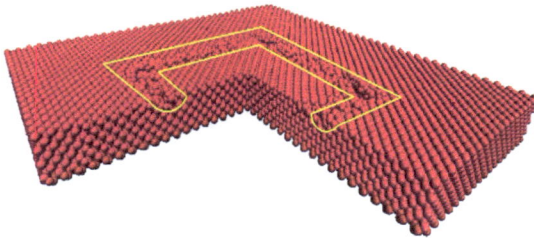

Figure 10: Cross section view of microstructure.   **Figure 11:** Geometry model of materials removal.

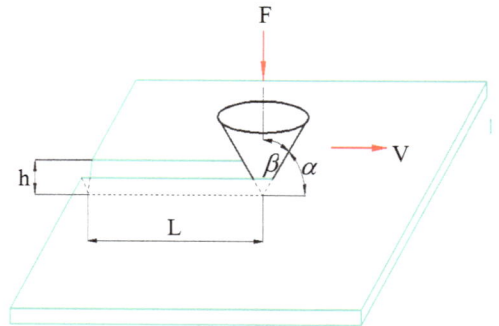

In fact, the removed atoms (213) computed by equation (2.57) and (2.58) is far beyond the real value (51) which means that the structure rearrangement induced by plastic deformation at this atomic scale dominate the substrate nanolithography process. There are not obvious chip formed in the whole process except at the start/end point where the AFM indent or lift from the substrate and the adhesive wear dominate at this stage. The residual depth of the groove is minors comparing with the original scratch depth which justify that AFM based nanolithography would be little-wear friction at the small depth of scratch (less than 10 nanometers). At the same time the lateral deformation is larger which makes the final mark similar as welding seam as shown in Fig. (**12**).

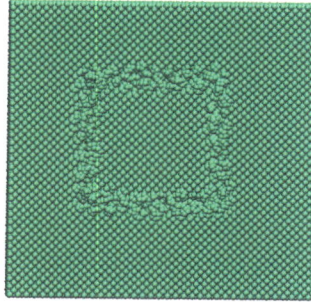

Figure 12: Residual impression.

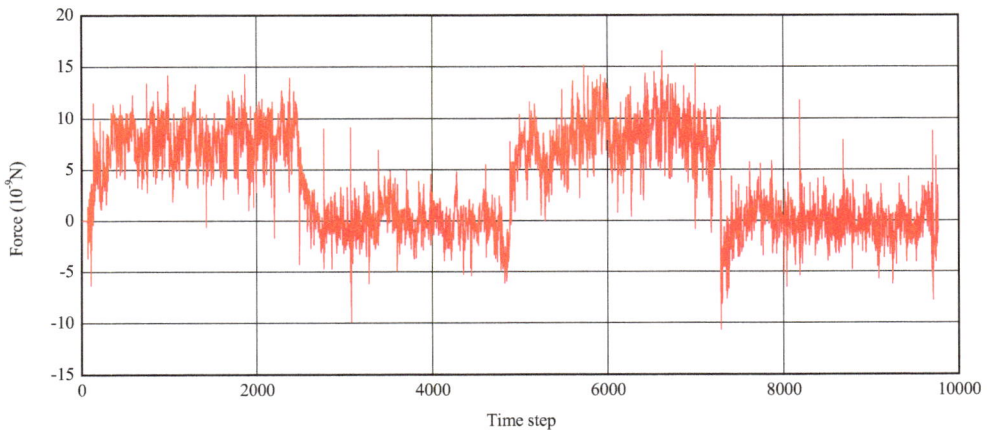

Figure 13: Variation tendency of force.

The little-wear tribology behavior could result in local disorder of crystal structure or amorphous layer which can change the materials electrical/magnetic property. The final micro-structure acquired by AFM based technique can only has the similar shape with the concept design because of the self-organization property of nano-materials. Also, the effect of this AFM based nanolithography does not extend far beyond the surface which keeps the original materials atomic structure. The mechanisms and dynamics of the interactions of two contacting solids during relative motion, ranging from atomic to microscale, need to be understood in order to develop fundamental understanding of AFM based nanolithography processes. At most solid–solid interfaces of technological relevance, contact occurs at many asperities. Consequently the importance of investigating asperity contacts in studies of the fundamental micro/nano-mechanical and micro/nano-tribological properties of surfaces and interfaces has long been recognized. During scratching, the tip moves discontinuously over the substrate surface and jumps with discrete steps from one potential minimum (well) to the next. This leads to a saw-tooth-like pattern for the lateral motion (force) with a periodicity of the lattice constant (Fig. (**13**)). This motion is called stick-slip movement of the tip. As just mentioned, stick-slip on the atomic scale is the result of the energy barrier that must be overcome to jump over the atomic corrugations on the sample surface. This corresponds to the energy required for the jump of the tip from a stable equilibrium position on the surface into a neighboring position. The perfect atomic regularity of the surface guarantees the periodicity of the lateral force signal, independent of actual atomic structure of tip apex. Few atoms (based on the magnitude of the friction force) on a tip sliding over an array of atoms on the sample are expected to go through the stick–slip. As the sample surface slides against the AFM tip, the tip remains stuck initially until it can overcome the energy (potential) barrier, which is illustrated by some definite kinds of interaction potential experienced by the tip. After some motion, there is enough energy stored in the spring which leads to slip into the neighboring stable equilibrium position. During the slip and before attaining stable equilibrium, stored energy is converted into vibrational energy of the surface atoms in the range of 1013 Hz (phonon generation) and decays within the range of 1011 s into heat. The stick–slip phenomenon, resulting from irreversible atomic jumps, can be theoretically modeled with discrete mechanical models.

5.3. Chemical Mechanical Polishing

Presently, the interconnect delay has already become an increasingly larger portion of the ultra-large scale integrated circuits delay as the device dimensions decrease. The interconnect delay is an important limiting factor which can be decreased by the use of lower resistivity metal and lower dielectric constant interlayer materials. Besides the electrical performance, other properties such as process ability (to form both *via* and interconnects), mechanical stability, reliability of a metal dictates its applicability as interconnect materials. Aluminum and its alloy have been used the material of choice for interconnect system since the beginning of integrated circuit because of its good conducting properties and compatibility with the fabrication devices. But Al/SiO_2 has relatively high resistance and dielectric constant which can contribute to the RC propagation delay. As feature size moves towards smaller geometries in order to achieve high speed and integration density, Al/SiO_2 interconnect system has found to be a limiting factor. Among metals for IC metallization, only silver, copper and gold have resistivity lower than aluminum. Copper has good resistivity characteristic and offers good electro-migration resistance. A significant improvement has been achieved by replacing Al interconnect with Cu. Cu/low-k interconnect technology has become a critical technology for 0.1 μm and sub-0.1μm device manufacturing due to the increased device speed, enhanced electro-migration resistance and improved scalability it affords.

With an increasing demand for high function and high performance ultra-large scale integrated circuits, chemical mechanical polishing (CMP), a final finishing process of the fabrication, has already been recognized as the most critical step throughout the process. For electronic components, with a growing tendency towards high density, extreme importance has been placed on the introduction of technologies to produce accurate topography and shapes in the order of 10^{-3} meter (micrometer) to 10^{-9} meter (nanometer). It is also indispensable from viewpoints of material properties to produce a completely smooth flat surface. By creating such high precision and high quality surfaces, the characteristics inherent to each functional material will be made available for full use, leading to the realization of high performance components. Such requirement has already approached the limit of manufacturing technology. Under this circumstance, any change of the flow field or conflict in physical-chemical factors can deteriorate the surface quality, any tiny hard particle may generate large pits or mark in the surface. Further study on the physical-chemical aspects of CMP is crucial for obtaining planarization surface with nanometer level roughness.

Despite intense theoretical and experimental research on CMP, there is still serious lack of fundamental understanding on this process. The application of CMP still rests on the semi-empirical stage. Presently, the researchers can't give convinced illumination about the mechanism of CMP. The reason for this is that the CMP process is a complex system characterized by multiphase, multi-scale and multilevel, at the same time being a micro/nano-tribology behavior based chemical physical process. It is not the geometrical downsizing but should exist some new discipline dominates the material remove and surface generation process in the CMP technique.

The planarization process was done primarily to remove surface damaged layers created by the previous process, and to achieve a specified wafer thickness and surface planarization. The damage may extend for several microns into the materials, and thus the planarization process had to be rapid, yet maintain a planar surface. The planarization process is viewed as being a chemical softening of the materials surface and the mechanical scraping of this softened layer by the abrasive particle. CMP is a technology that accurately produces geometrically dimensional shapes in the nanometer order. Fine abrasive particles are retained on the pad surface resiliently and plastically, and the work surfaces are scratched microscopically. Planarization actions are by far smaller if compared with lapping, contributing to the successful applications to the brittle materials such as single crystal silicon. The final surface integrity acquired using CMP technique is mainly depended on the micro-tribology behavior of nanoparticles. The complex multi-body interaction among nanoparticles and materials surface is different from interaction in the macroscopic scale which makes the traditional classical materials machining theory can't accurately uncover the mystery of the surface generation in the CMP.

In reality, planarization may occur by either solid-based or by fluid-based wear. During solid-based wear, the abrasive particles are dragged across the surface and act as cutting tools. Removal volumes are determined by abrasive particle loading and film properties. During fluid-based wear, abrasive particles are not dragged across the surface, but rather impinge on the surface at some velocity and angle. As particles collide with the surface, they impart energy to the surface, resulting in strain, weakened bonds, and eventually material removal. Planarization mode and the role of the fluid layer are poorly understood at this point. It is clear, however, that planarization mode and fluid layer thickness and continuity have important implications for polish rates and planarity. This area of CMP is still poorly understood, yet has important implications as to the removal mechanisms of CMP. The traditional model for sliding wear such as the *Archard* wear equation [8] cannot give a convinced explanation about this discrete particle tribology behavior.

Interaction between nanoparticle and solid surface are of fundamental importance to science and technology. As the size of the particles decreases and approaches molecular dimensions, their atomic nature becomes increasingly important. Interactions on the molecular level may be studied experimentally using molecular beam techniques, and theoretically by molecular dynamics simulations. In the current paper (molecular dynamics) MD simulations are employed to study the interactions between copper nanoparticle and substrate. Nanoparticles are relatively large and may be expected to behave as "macroscopic" particles in some respects. At the same time they are small enough to permit the long-time simulations necessary for investigation of self-structural organization processes. Interactions between nanoparticles and surfaces have now received considerable attention for more than a decade due to their relevance for both fundamental and applied research. A variety of processes may occur depending on the properties of the nanoparticle, the available energy and the energy transfer rates involved. A straightforward interpretation of experimental results is often not possible due to the complexity of the dynamics. The use of MD simulation techniques has therefore become a very important tool in this field.

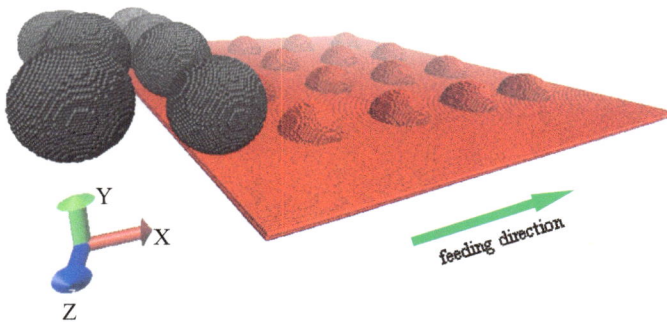

Figure 14: MD model on chemical mechanical polishing.

The atomic configuration of the system studied is illustrated in Fig. (**14**). Copper atoms of slurry particle and substrate are initially arranged in face-centered cubic structure with a constant lattice parameter of 3.615 Å which are uniformly distributed. The dimension of copper workpiece is $200 \times 1 \times 200 \, \text{Å}^3$ along x, y and z directions. The moving speed is 20 m/s, the environment temperature is 293 K, the particle diameter is 20 Å, the timestep is 1.5×10^{-15} s. In this study, the phenomenological model includes 16 asperities (hemispheres) and 8 particles (spheres) which are uniformly distributed in $350 \times 200 \, \text{Å}^2$ area. The asperities are considered to be on top of the thin film such that the center of each asperity lies in the plane of film surface and each asperity is hemispherical. The abrasives are assumed to be spherical particles. The EAM potential is used to depict the interaction between the copper atoms. The Velocity-Verlet method is a symplectic algorithm which can prevent the energy dissipation and have high computation efficiency and being adopted in this investigation.

Fluid-based wear in the CMP (Fig. (**15**)) is caused by hard particles that are free to roll or slide between two sliding surfaces (polishing pad and substrate) which is also termed as three-body abrasion. Despite the technological importance of wear, no simple and universal model has been developed to describe it. As with many other tribological phenomena, the multitude of physical mechanisms contributing to wear makes it difficult to develop a general comprehension of how wear occurs at the nanoscale. As the size of the particles decreases and approaches molecular

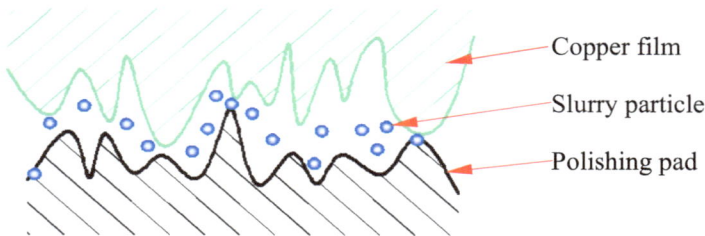

Figure 15: CMP working conditions.

dimensions, their atomic nature becomes increasingly important. Nanoparticles are relatively large and may be expected to behave as "macroscopic" particles in some respects. At the same time they are small enough to permit the long-time simulations necessary for investigation of structural self-organization processes. A variety of processes may occur depending on the properties of the nanoparticle, the available energy and the energy transfer rates involved. A straightforward interpretation of experimental results is often not possible due to the complexity of the dynamics. Understanding the micro-mechanisms contributing to three-body abrasion and their microdynamic behavior is crucial for exploring fundamentals of CMP.

The simulation starts with a consideration of a small number of abrasives and asperities on copper thin films. The abrasives are allowed to move linearly with constant velocity. Periodic boundary conditions (along x direction and y direction) are used to shift abrasive from one side to the other side of region. Fig. (**16**) gives the MD simulation results on complex interaction among nanoparticles and single crystal copper film and also the particles tribology behavior. Firstly, these particles move parallel to the x direction while the rotational movement induced in the planarization process is neglected. The trajectory of all particles remain along x axis which means the impact between particles and surface peak are normal impact. The tool (particles) should keep its dynamic behavior (micro & nanoscale inertia) between two discrete planarizing events if there is not external load exerted on it. As there is no rigid constraint which makes the tools gradually deviating from their original track, the tool and substrate keep away from each other little by little. The particles are gradually lifted after impacting the substrate materials. This dynamic trajectory leads to the first peak adjacent to the particles being planarized sufficiently while other following peaks being planarized unsufficiently. Most of the materials removed from the substrate stick to the particle surface which means this materials removal process is mainly generated by adhesive wear which exceeds the yield stress of the sliding materials. The notion that wear should be related to adhesion seems fairly natural: the same bonding mechanism that makes it difficult to pull surfaces apart should also make it difficult to slide them over each other. The adhesion occurs after the particles touch the surface peak, followed by plastic shearing that plucks off part of the asperities; these bits then adhere to the outer layer of the particles before eventually becoming loose wear debris. The planarization process can be considered as a process with energy release. A rough surface has high energy and a smooth surface has low energy. Since adhesive forces scale with surface energy, the initial rough surface with higher surface energy should result in higher wear rates.

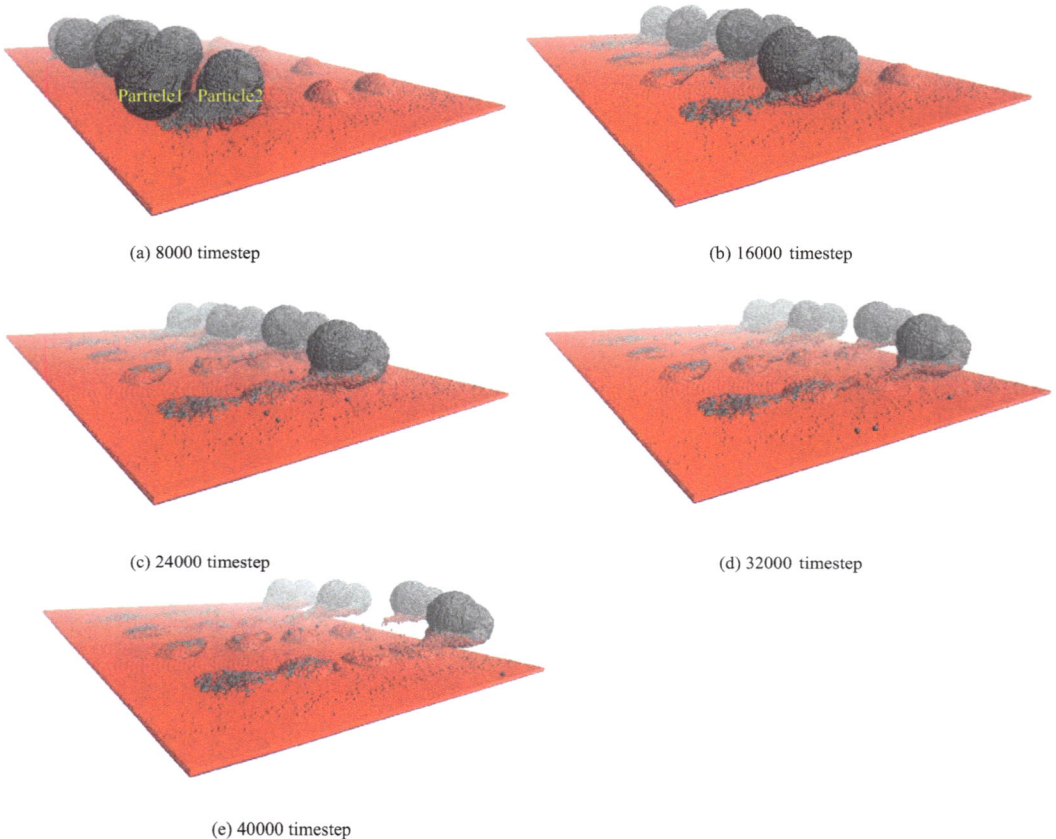

(a) 8000 timestep

(b) 16000 timestep

(c) 24000 timestep

(d) 32000 timestep

(e) 40000 timestep

Figure 16: MD simulation results about CMP.

The interaction potential energy between particle1 and particle2 decreases after the initialization of CMP process and finally tends to be stable (Fig. (**17**)). The pair interaction gradually decreases accompanying the rearrangement of atomic structures. This graph justifies the generation of new particle clusters in the CMP process. The similar variation trend is observed between other particles. Furthermore, the dynamic behavior of the two particles is influenced by the interaction between particles and the substrate which affect the surface planarization. This kind of substrate-particle interaction act as integrate motion constraint because the substrate is fixed. It is just this particle relative kinematic motion realizing the surface planarization. The particle-particle interaction belongs to non-integrate motion constraint because both of the two particles also move independently. The interaction force between the two particles becomes stronger as they come closer and finally a larger atomic cluster formed which makes the two particles moving like a unity. The linkage between these particles belongs to internal flexible force field which is different from the macroscopic tools such as grinding wheel. The physical collision between particle and the substrate should be some elementary kinds of chemical reaction as there are chemical bond fractured and new particle cluster generated. The system relax fully before the first collision, the substrate surface is in some comparatively stable state. The substrate surface atom become chemical active after pair collision, its activation barrier decreased and the reaction rate increased.

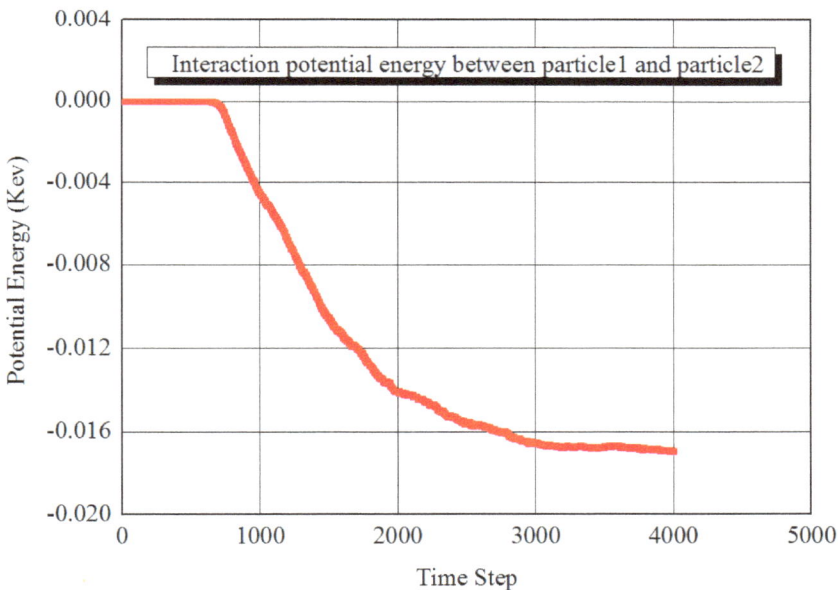

Figure 17: Interaction potential between two particles.

Fig. (**18**) shows the surface morphology after CMP. As both of the slurry particle and the surface peaks are simplified as sphere, the residual finished surface (Fig. (**18 (a)**)) can be assumed as shown in Fig. (**18 (b)**). This result justifies that the real finished nanometer level surface should be the mathematical integration of

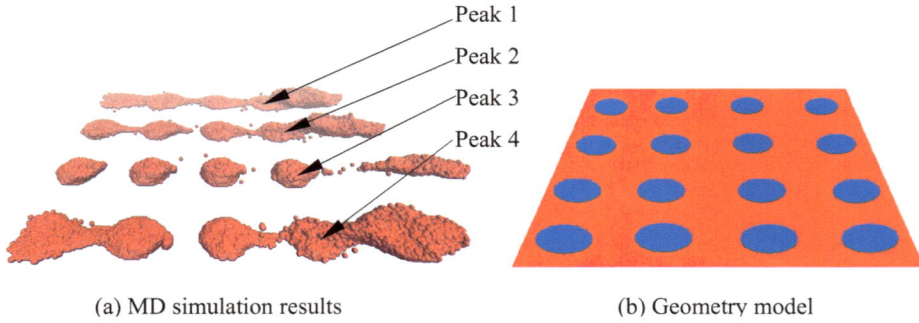

(a) MD simulation results   (b) Geometry model

Figure 18: The residual impression after CMP.

numerous discrete areas but not a continuous smooth integrity in the view point of atomic level. Strong shear interaction (Fh) will result in lateral spreading while weak shear interaction will generate water drop shape residual area. Furthermore, the author's previous research together with this work (Fig. (**19**)) justify that the surface planarization effect would be decreased with the feeding of particles as there are not rigid constraints which can restrict the motion of suspension particles and the wear of the grain. The final surface should be the repeated synergic behavior of all particles using various processing technology such as cutting, impacting, scratching, indentation and so on. Fig. (**19**) gives the evolvement of the interaction force vector between particle and surface peak, where F is the total force, Fh is the horizontal force component, Fv is the vertical force component.

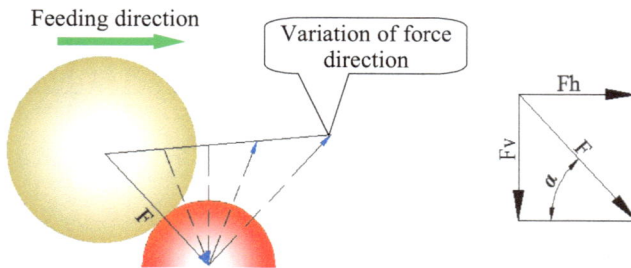

Figure 19: The collision between two particles.

As the slurry particle gradually approaching the surface peak, namely, the slurry particle will compress the surface peak if the α less than 90°. On the other hand, the slurry particle will drag the surface peak if the α great than 90°. This analysis can be justified by the MD simulation results showed in Fig. (**16**). The Fh component or the shear effect will be strengthened greatly if the α is small which will improve the planarization effect as shown in the Fig. (**16**). The impact angle (α) of peak 1 and peak 4 are less than that of peak 2 and peak 3 which result in better surface planarization effect and high efficient materials removal.

CMP is a multibody nanometer machining technology, and there exists complex interaction among particles as well as the interaction among particles and substrate materials. As this interaction falls into the nanometer regime, it is not downsizing but some new discipline and mechanism dominate the CMP technology. Planarization may occur by either solid-based or by fluid-based wear. During fluid-based wear, abrasive particles are not dragged across the surface, but rather impinge on the surface at some velocity and angle. As particles collide with the surface, they impart energy to the surface, resulting in strain, weakened bonds, and eventually material removal. The collisions between abrasive particles and the pad accelerate the abrasive particles. The particles then impinge on the wafer surface, resulting in fluid-based wear. In addition because the particles lose energy (slow down) as they move through the fluid layer, local fluid layer thickness and slurry viscosity will also affect the particle velocity. Planarization mode and the role of the fluid layer are poorly understood at this point. This area of CMP is still poorly understood, yet has important implications as to the removal mechanisms of CMP. The complex multi-body interaction among suspending nanoparticles and crystal wafer is crucial for uncover the mystery of the materials removal and surface defect generation in the CMP. The authors investigate CMP process of single crystal silicon using MD simulation method, after that drew some conclusions:

(1) The velocity and angle of approach of the abrasive particles will determine the kinetic energy that the particles transfer to the surface, and hence will affect removal rates.

(2) The physical collision between particle and the substrate should be some elementary kinds of chemical reaction as there are chemical bond fractured and new particle cluster generated.

(3) The real finished nanometric surface should be the mathematical integration of numerous discrete areas but not a continuous smooth integrity at atomic level.

5.4. Nanoindentation

Recent developments in the field of nanotechnology and nanofabrication have produced material structures with nanometer scale dimensions. Thin film is one of the fast growing sciences whose length scale lies in this domain. It plays crucial role in the high-tech industries that will bear the main burden of future technological developments. Thin films often support very high stresses during services which can lead to the distortion of the devices containing the film. Although many materials are selected for their electrical, magnetic, or optical properties, it is often the mechanical properties that limited the fabrication and reliability of many potential advanced devices. Therefore, knowledge of the mechanical properties of thin films is critical for the design, fabrication and application of advanced devices. The practical application of these devices requires a thorough understanding of their mechanical properties on the nanometer scale. The principal goal of such mechanical testing is to obtain values for elastic modulus and hardness of the specimen material from experimental readings of indenter load and depth of penetration. The forces involved are usually in the milli-newton range and are measured with a resolution of a few nano-newton. The depths of penetration are in the order of nanometers, hence the term "nanoindentation". Investigation of the nanoindentation can provide the knowledge of surface contact nanomechanics required in the nanoscale wear processes, such as those occurring in materials manufacturing processes, and nanofabrication technologies such as the AFM-based data storage technology, and nanoimprint lithography and MEMS. Nanoindentation has proven to be the simplest and most direct way to determine the mechanical properties of thin films.

Experimental procedure is appropriate of investigating the bulk materials mechanical property in the larger length scale. However, microhardness testing instruments available could not apply low enough forces to give penetration depths less than the required 10% of the film thickness so as to avoid influence on the hardness measurement from the presence of the substrate. Even if they could, the resulting size of the residual impression cannot be determined with sufficient accuracy to be useful. For example, the uncertainty in a measurement of a 5 μm diagonal of a residual impression made by a Vickers indenter is on the order of 20% when using an optical method and increases with decreasing size of indentation and can be as high as 100% for a 1 μm impression. Therefore, determining the intrinsic properties of the films can be difficult because of the influence of the substrate on measured properties. Since the spatial dimensions of the contact area are not conveniently measured, modern nanoindentation techniques typically use the measured depth of penetration of the indenter and the known geometry of the indenter to determine the contact area. For such a measurement to be made, the depth measurement system needs to be referenced to the specimen surface, and this is usually done by bringing the indenter into contact with the surface with a very small "initial contact force," which, in turn, results in an inevitable initial penetration of the surface by the indenter that must be accounted for in the analysis. Additional corrections are required to account for irregularities in the shape of the indenter, deflection of the loading frame, and piling-up of material around the indenter. These effects contribute to errors in the recorded depths and, subsequently, the hardness and modulus determinations. On the other hand, we need to recognize that material properties measured by nanoindentation are generally evaluated according to contact mechanics of isotropic elastic bodies. For instance, if the material under investigation is indented by a spherical indenter, Hertzian elastic solution [9] is applied to evaluate its elastic properties. This approach is widely used and yields reasonable results in certain cases. However, the assumption of isotropy may lead to critical errors in the case of metals and materials with strongly anisotropic structures.

Furthermore, the scale of deformation in nanoindentation test becomes comparable to the size of material defects such as dislocations and grain sizes, and the continuum approximation used in the analysis can become less valid. There is no doubt that as the scale of mechanisms becomes smaller, the nature of surface forces and adhesion will continue to increase. Therefore, a powerful theoretical toolkit which can describe the microscopic world effectively is needed to investigate the physical property of thin film at small scale. Molecular dynamics (MD) simulation, by virtue of its high temporal and spatial resolution, can offer an ideal approach to gain insights into atomic scale process and understand their mechanisms. This research aims to verify the effects of size scales on the physical property of copper films using numerical simulation method. The observed numerical experimental trends exhibit a strong size dependence, which cannot be rationalized on the basis of continuum concepts. Computer simulation of the nanoindentation using molecular dynamics (MD) is conducted to elucidate the mechanism about elastic-plastic behavior and anisotropy property. By correlating observations from computations, physical mechanisms responsible for those are discussed.

The atomic configuration of the studied system is illustrated in Fig. (**20**). The diamond indenter is assumed as ideal rigid body without any deformation. Copper atoms of workpiece are initially arranged in fcc structure with a constant lattice parameter of 3.61 Å. The dimension of copper workpiece is $100 \times 80 \times 100 \, Å^3$ along x, y and z directions. The moving speed is 30 m/s, the environment temperature is 293 K. The indenter radius is 15 Å, the indent depth is 15 Å. The EAM potential is used to depict the interaction between the copper atoms. The interaction between the copper atom and the diamond atom is modeled by the Morse Potential. The Velocity-Verlet method is a symplectic algorithm which can prevent the energy dissipation and have high computation efficiency and thus being adopted in this investigation.

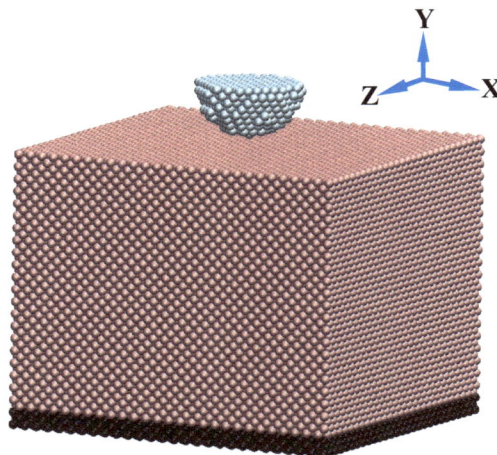

Figure 20: MD Simulation Model.

In this simulation, a sharp rigid diamond indenter is initially positioned above the surface of the thin film. The indenter is moved downward incrementally to indent thin film. Once the indenter has reached the specified indentation depth, it is retracted from the thin film and returned to its original position. At each incremental displacement of the indenter, the force equilibrium equations of the atoms as determined by their inter-atomic potentials are nonlinear functions of their incremental displacements. These equations are solved using the Newton-

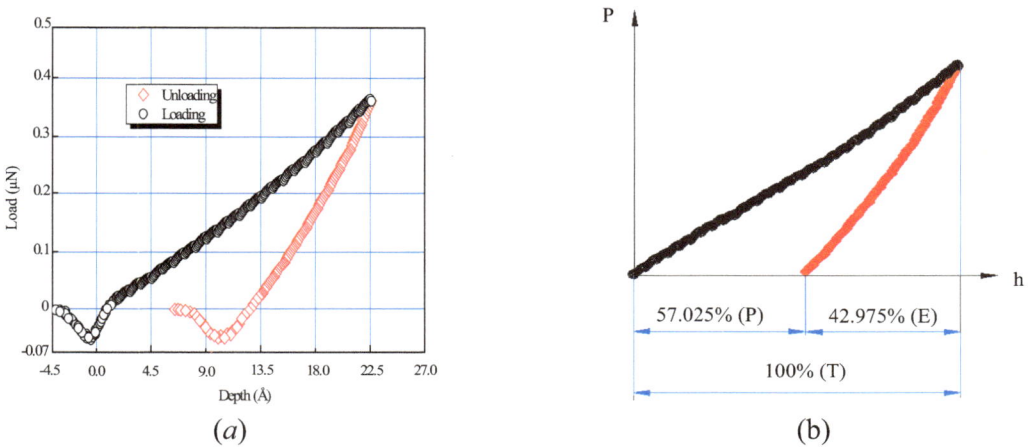

(a)   (b)

Figure 21: MD simulated *P–h* response of 8-nm-thick Cu films.

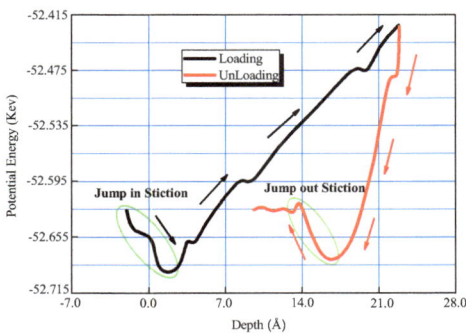

Figure 22: MD simulated copper substrate potential in the indentation process.

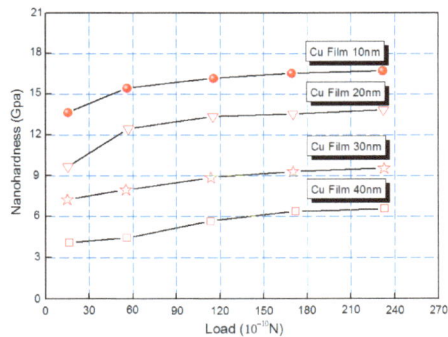

Figure 23: Indentation size effect in the nanoindentation Cu films with different thickness (MD simulated results).

Raphson iterative technique. Figs. (**21-27**) gives the numerical simulation results of indentation single crystal copper film. Fig. (**21**) indicates the relationship between the load, *i.e.*, the force experienced by the indenter, and the indentation depth over the complete nanoindentation cycle. As the rigid indenter approaches substrate, surface atoms firstly jump into contact with the tip and the interfacial force is attractive. The adhesion force reaches its maximal value after the indenter move downward 4.157Å (approximately 0.0518μN). The length of the chemical bond basically keeps constant indicates that the adhesion-contact status induced at the interface. The length of the chemical bond basically keeps constant indicates that the adhesion-contact status induced at the interface. The contact load increased quickly after the indenter penetrating through the surface of copper film

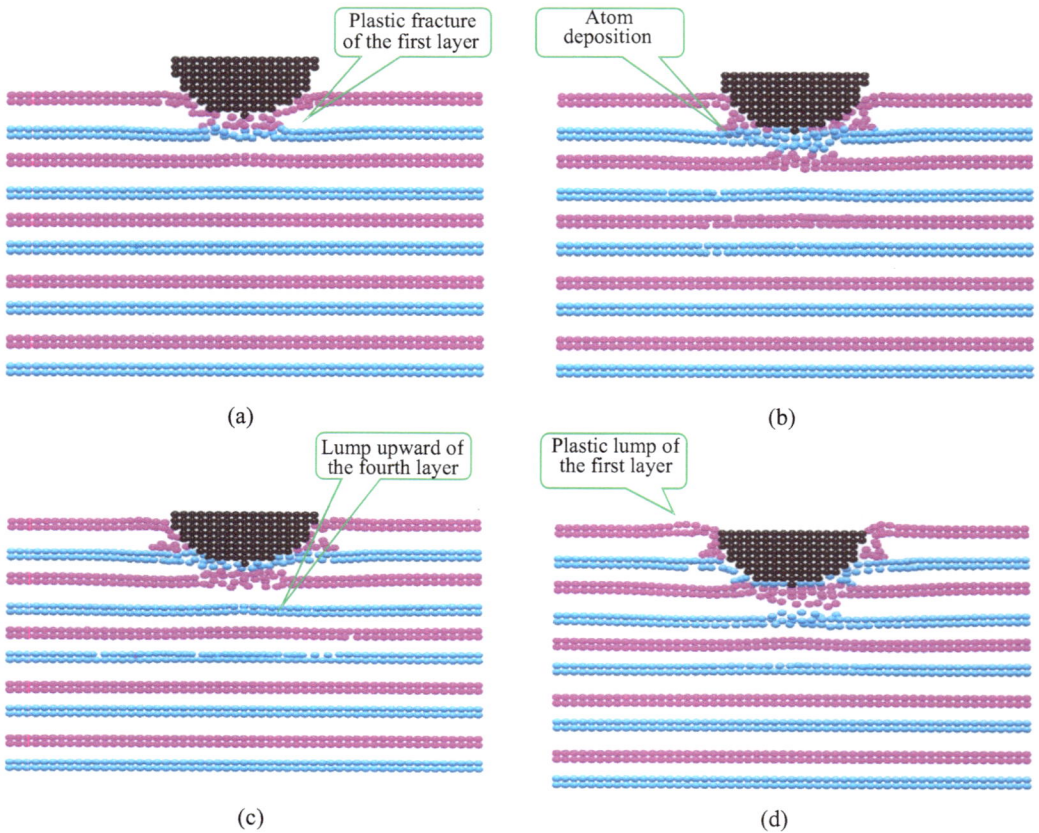

Figure 24: Layer penetrated by indenter at different stage about MD simulated nanoindentation of Cu.

and the atomic adhesion behavior is initialized as shown in Fig. (**21**). At this micrometer scale, material behavior is mainly controlled by surface-driving effects than by bulk effects. Surface contact of micromechanical structures can contribute to stiction (adhesive contact), damage, pitting, and surface hardening over local contact area. The presence of stiction is due to high compliance and large surface-to-volume ratio of micromechanical structures, which do not have enough restoring force to overcome the surface interaction after mechanical contact. Stiction can be classified into two categories: jump in stiction and jump out stiction (Fig. (**22**)). There are opposite evolvement tendency about the system potential energy under the two stiction conditions. The potential energy gradually decreased in the jump in stage as the indenter doing negative work with the copper film while the indenter doing positive work with the copper film in the jump out stage and result in the increasing of system potential energy. The force response during loading exhibits some fluctuations due to the inducing of complex dislocation in the materials. Another notable feature is the decaying of the load with time at maximum displacement which causes the slope of the load-displacement curve diverging there. This phenomenon justifies the elastic-plastic deformation property of copper film in the nanometer scale and there must be plastic deformation energy stored in the copper film. The surface atoms gradually lump upward as the system temperature increased from 293 K to 427 K. After that, the temperature of the contact area gradually drops to 300 K as the heat dissipated away by means of random colliding.

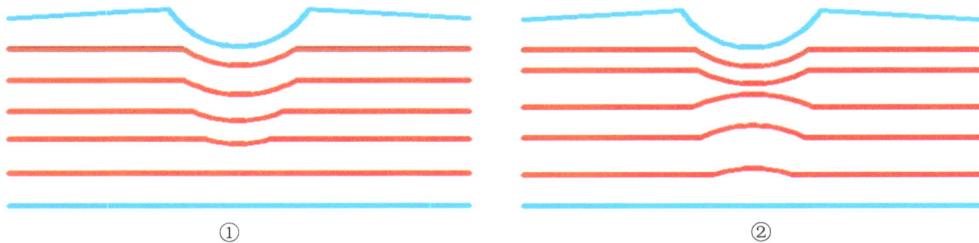

Figure 25: Layer deformation induced by indenter (displacement model) ② Finite element results; ② Molecular Dynamics results.

In the nanometer or sub-nanometer scale, it is the surface force that dominates materials physical property which makes their micro-mechanical property different from corresponding macro-mechanical property. While in the unloading process, there are still attractive force existed after the indenter retracted from the deformed copper thin film. The transformation of control mechanism in this small length scale (such as surface force and thermal fluctuating) may exert great effect upon materials micro-hardness. Fig. (**21**) shows the load-displacement responsive property under the nanometer scale where the elastic deformation occupies 42.975 percent of the entire deformation in this atomic length scale. This data justifies that the permanent plastic deformation quickly decreased while the elastic deformation gradually increased in the range of nanometer level. The decreasing of residual deformation (plastic deformation) means the increasing of the material's micro-hardness.

The hardness of the single crystal copper along (010) plane calculated from MD simulation results approximately to be 12.3 GPa while the corresponding hardness calculated from finite element simulated results approximately to be 1.483 GPa. At the same time, Fig. (**22**) shows that the nanohardness of copper film increased with the decreasing of film thickness. These two results together justify that there exists size effect in the materials mechanical property, namely, the nanohardness of copper film is greater than its corresponding macrohardness. Furthermore, the nanohardness of copper film gradually increases accompanying the increasing of indentation load. This phenomenon illuminates there are indentation hardening behavior in the nanoindentation of thin film materials. There may be several factors which can account for this phenomenon. Firstly, the probability of encountering stress-reducing defect decreases as the size of the material decreases. Secondly, the elastic recovery after removal of indenter will lead to the decrease in the residual indentation size which will lead to the increase of computation materials hardness. Smaller samples are expected to have less free volume than the bulk materials, therefore if the volumes are small enough, shear band propagation, which is a primary mechanism of plasticity in the metal might not even be possible. Plastic deformation confined within a very small volume underneath the indenter may result in the creation of non-uniform stresses and strains within the sample, thereby setting up strong gradients of strain, which are also to be one of the responsible factors for the larger materials nanohardness.

The deformation condition inside the film is showed in Fig. (**24-25**). In the Fig. (**24**), every other layer of the substrate materials are revealed for clarity. Firstly, the top layer is elongated and gradually resulted in plastic fracture. After that, part of atoms belongs to this layer depositing on the adjacent layer. The workpiece surface gradually sunk when the indenter keeps feeding and result in the downward protuberant of atomic layer. This trend terminates at somewhere beneath the material surface (shown in Fig. (**24-25**)). The protuberant behavior in the opposite direction beneath the surface may be the effect of stiction between the adjacent layers. This abnormal phenomenon should have tight relationship with the length scale in the indentation direction. The different deformation behavior of crystal layers inside the films may be the potential key factor of its breaking off from the substrate. This phenomenon should be a collective behavior, namely, all of the atom at the same layer, but not any discrete atom behavior. It is justified by many researchers (also include the author) that the ductile material surface will lump upward when the indenter approaches workpiece and thus influence the depth of indentation. The animation movie of the MD simulation result shows that there are still stiction effect beneath the material surface in the nanoindentation process. Comparing with the macro-indentation process, this phenomenon should be directly resulted from size effect and have great effect upon materials micro-mechanical property.

The atom structure at the local indentation region is shown in Fig. (**26**). In this figure, only the local deformation substrate atoms are revealed. The perfectly coordinated atoms (N=12) far away from the indentation region have been removed from the snapshots for clarity, such that only surface and imperfectly coordinated atoms are visible. We notice that in the case of the indentation, huge pressure being exerted on the local contact area and result in the increasing of atom coordinate numbers (13 and 14). Most of the atoms in the normal region hold 12 neighbors while this value gradually increased to 14 at the symmetrical center of the workpiece (intense deformation area). This result justifies that the material structure in this region undergone solid-state amorphization and the film hardening behavior just originates from the rearrangement of crystal structure at the atomic level.

Fig. (**27-29**) give the numerical simulation results of anisotropy about crystal copper from nanoindentation. There are basically two aspects which can induce the anisotropy phenomenon, namely, the indenter and the substrate. The spherical

Coordinate numbers

- 14
- 13
- 12

(a) 1500 timestep

(b) 3000 timestep

(c) 4500 timestep

(d) 6000 timestep

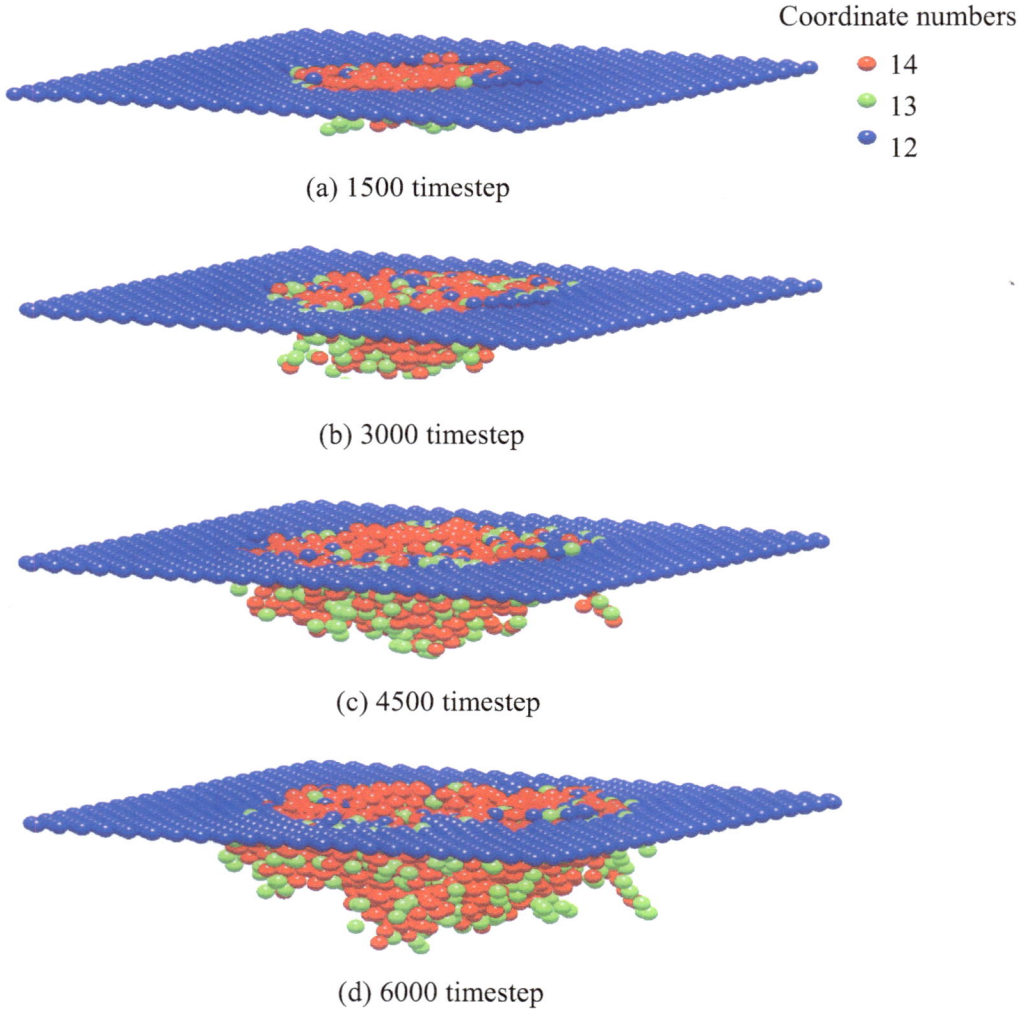

Figure 26: Number of atoms with specified nearest number of neighbors at different stage of MD simulated nanoindentation of Cu.

indenter is adopted to eliminate the influence of the indenter geometry. Therefore, this anisotropy phenomenon should be entirely induced by the physical property of the substrate and belongs to planar mechanics behavior as the thickness of the film can be ignored comparing with the other two dimensions of the film. This anisotropy of the thin film should have different essence as the its special physical structure may influence the elastic-plastic deformation behavior. Presently, there are two kinds of anisotropy, namely, structure anisotropy and texture anisotropy.

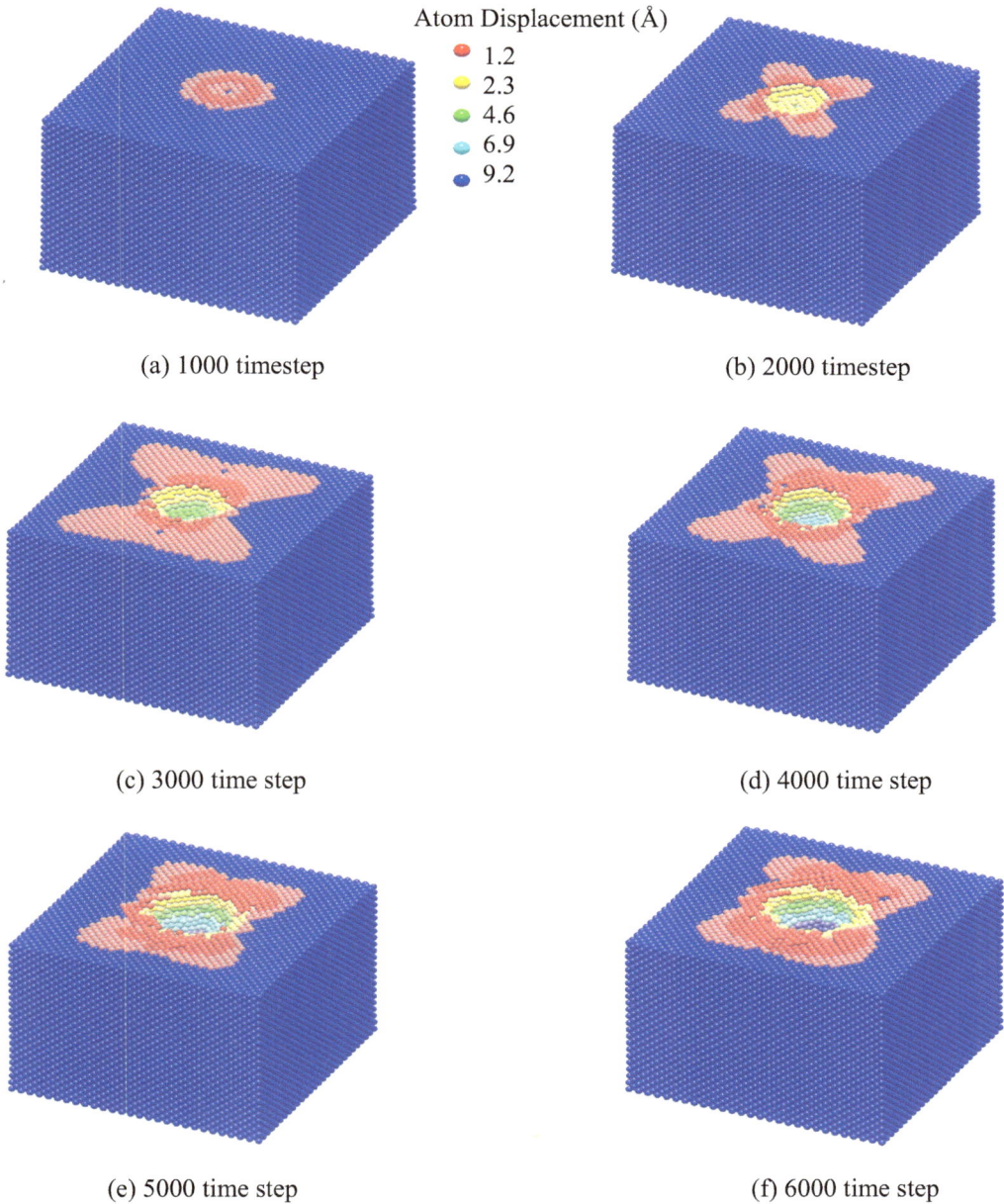

(a) 1000 timestep

(b) 2000 timestep

(c) 3000 time step

(d) 4000 time step

(e) 5000 time step

(f) 6000 time step

Figure 27: The atom displacement from MD simulated nanoindentation results.

The anisotropy deformation in the thin film belongs to structure anisotropy as the atomic arrangement in the crystal materials has its definite order. As showed in the simulations, material from the damaged region was essentially non-contiguous, irregularly shaped lumps at several direction illustrated in Fig. (**27**). Initially, the contact between two bodies belongs to the elastic Hertz contact [9] and the deformation is identical in every direction. Gradually, four sets of separated parts ("handles") generated in the in-plane [101] directions. This results justify the strong anisotropy deformation exists in the copper film. The anisotropy deformation has nothing to do with the materials inhomogeneities as the structure and the property are identical at any point of crystal interior. The anisotropy also has strict preferred orientation distribution discipline (Fig. (**28**)). Firstly, this anisotropy deformation of surface is symmetrical distribution about (100) and (001) plane. Secondly, there is more intense plastic deformation generated in the in-plane [101] directions than in any other direction. Finally, the anisotropy deformation gradually minimized accompanying the increasing of plastic deformation. As the copper is a typical elastic-plastic materials which makes there are only local surface lumping instead of materials pileup in the indentation of rigid-plastic workpiece. The elasticity of single crystal copper plays an important part in the plastic nanoindentation process. When the yield point is first exceeded the plastic zone is small and fully contained by material remains elastic so that the plastic strains are of the same order of magnitude as the surrounding elastic strains. In these circumstances the material displaced by the indenter is accommodated by an elastic expansion of the surrounding solid.

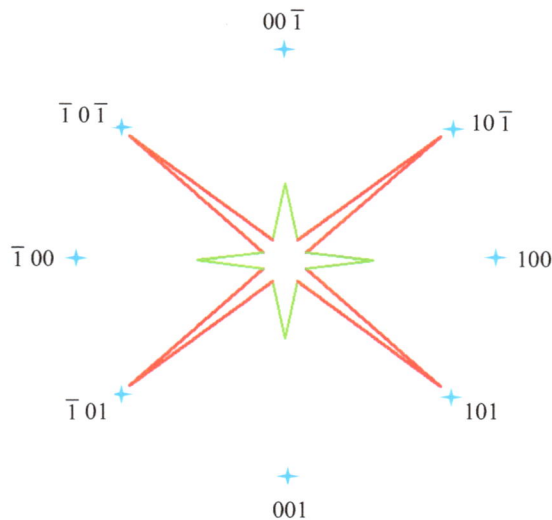

Figure 28: The anisotropy property projected on the (010) plane.

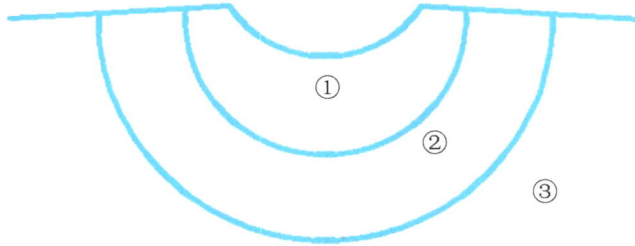

Figure 29: Impression cross section (① amorphous region; ② plastic deformation; ③ elastic region.

As the indentation becomes more severe, either by increasing the load on a curved indenter or by using a more acute-angled wedge or cone, an increasing pressure is required beneath the indenter to produce the necessary expansion. Eventually the plastic zone breaks out to the free surface and the displaced material is free to escape by plastic flow to the sides of the indenter and resulting in the materials lump (Fig. (**29**)).

The determination of mechanical properties of nanoscale structure is becoming increasingly important as microsystem and ultra-large scale integrated circuit technologies continue to mature. Traditional experimental methods can't avoid the influence on the hardness measurement from the presence of the substrate. Numerical computation method of finite element modeling with molecular dynamics simulation is used to determine the mechanical properties of copper thin film from indentation, quantifying the difference between load *versus* displacement-into-surface curves obtained from different length scale. The results show that the materials deformation exhibits strong size dependence when the relevant physical length scales fall into the range of microns or below. The P-H graph justifies that the permanent plastic deformation quickly decreased while the elastic deformation gradually increased in the range of nanometer level which means the increasing of material micro-hardness. The different deformation behavior of crystal layers inside the films may be the potential key factor of its breaking off from the substrate. There are little anisotropy phenomenon in the elastic deformation stage while obvious anisotropy phenomenon in the plastic deformation process. The anisotropy of thin film has strict preferred orientation distribution and symmetry. The anisotropy deformation gradually minimized accompanying the increasing of plastic deformation illuminates the copper thin film is a non-liner anisotropy elastic-plastic material.

Molecular dynamics simulation provides unique and powerful armament for MNT research on the atomistic model which could produce a clear picture of physical and chemical process and the foundation of accurate, robust and physically clear macroscopic models. However, molecular dynamics technique need potential function or empirical force field which can accurately describes the interaction between atoms. In essence different force field is required for each phase of a material which can enable MD to simulate phase transformation. Presently, the scale of MD simulation presented in this chapter is considerably less than experimental results and it is difficult to carry out direct compare. Nanometer scale model is small enough to preclude the formation of cracks of sufficient length to result in fracture although voids or crack do appear in the simulation results. Thoroughly investigation of many novel phenomena requires model size much larger than those presently in use. It may be continuum models more useful for studying phase transformation if accurate constitutive equations can be obtained to reflect the phase transformation process. It is necessary to develop energy functions based on quantum mechanical calculations to be used in the MD simulations. Future work ought to be directed towards the combination of MD technique with continuum mechanics finite element method to retain the atomistic description of the MNT process and allow the model to include macroscopic events which occur in the machining.

REFERENCES

[1]   M.P. Allen, D.J. Tildesley, *"Computer simulation of liquids"*. Oxford: Clarendon Press, 1987
[2]   J.M. Haile, *"Molecular dynamics simulation"*. New York: John Wiley & Sons, Inc, 1992
[3]   B. J. Alder, T.E. Wainwright, "Phase transition for a hard sphere system", *J. Chem. Phys.*, vol. **27**, pp. 1208-1209, August 1957.
[4]   A. Rahman, "Correlations in the Motion of Atoms in Liquid Argon", *Phys. Rev.*, vol. 136, pp. 405-411, May 1964.
[5]   F. H. Stilinger, A. Rahman, "Improved simulation of liquid water by molecular dynamics", *J. Chem. Phys.*, vol. **60**, pp. 1545-1558, September 1973.
[6]   M.S. Daw, M.I. Baskes, "Embedded-atom method: Derivation and application to impurities, surfaces, and other defects in metals", *Phys. Rev. B*, vol. 29, pp. 6443-6453, June 1984.

[7]   B. Bhushan, H. Fuchs, S. Kawata, "Applied scanning probe methods V". Berlin: Springer, 2007

[8]   J. F. Archard, "Contact and rubbing of flat surfaces", J. Appli. Phys., vol. 24(8), pp. 981-988, April 1956.

[9]   H. Hertz, "On the contact of elastic solids", J. Reine Angew. Math., vol. 92, pp. 156-171, March 1881.

Send Orders for Reprints to reprints@benthamscience.net

CHAPTER 3

Finite Element Simulation of MNT

Abstract: The dimensional tolerance achieved by precision machining technology is on the order of 1 nm and the surface roughness is on the order of 0.1 nm. The dimensions of the parts or elements of the parts produced may be as small as 1 μm, and the resolution and the repeatability of the machine used must be of the order of 1 nm (10 nm). Unlike conventional machining processes, precision machining processes are not based on the removing the metal in the form of chips using a wedge shaped tool. When metal is removed by machining there is substantial increasing in the specific energy required with decrease in chip size. Since the shear stress and strain in metal cutting is unusually high, discontinuous microcracks usually form on the metal-cutting shear plane. Owing to the complexity of elastic-plastic deformation at nanometer scale, the worldwide convinced precision materials removal theory is not built up until now. As the complexity associate with the precision machining process involve high strains, strain rates, size effects and temperature, various simplifications and idealizations are necessary and therefore important machining features such as the strain hardening, strain rate sensitivity, temperature dependence, chip formation and the chip-tool interface behaviors are not fully accounted for by the analytical methods. Experimental studies on precision machining are expensive and time consuming. Moreover, their results are valid only for the experimental conditions used and depend greatly on the accuracy of calibration of the experimental equipment and apparatus used. Advanced numerical techniques such as Finite Element Method (FEM) is a potential alternative for solving precision machining problems. Characterizing the surface, subsurface, and edge condition of machined features at the precision scale in the FEM analysis are of increasing importance for understanding, and controlling the manufacturing process.

Keywords: Finite element, thermal-mechanical coupling, materials removal criterion, boundary condition, residual stress.

INTRODUCTION

The need to manufacture high precision items and to machine difficult-to-cut materials led to the development of the newer machining processes. The dimensional tolerance achieved by precision machining technology is on the order of 1 nm and the surface roughness is on the order of 0.1 nm. The dimensions of the parts or elements of the parts produced may be as small as 1

Xuesong Han

μm, and the resolution and the repeatability of the machine used must be of the order of 1 nm (10 nm). The accuracy targets for ultra-precision component cannot be achieved by a simple extension of conventional machining processes and techniques. They are called precision machining processes, notwithstanding that the definition of conventional and traditional changes with time. Unlike conventional machining processes, precision machining processes are not based on the removing materials in the form of chips using a wedge shaped tool. There are a variety of ways by which the material may be removed in precision machining processes. Some of them are abrasion by abrasive particles, impact of water, thermal action, chemical action and so on.

When materials are removed by machining there is substantial increase in the specific energy required with decrease in chip size. It is generally believed this is due to the fact that all materials contain defects (grain boundaries, missing and impurity atoms, *etc.*), and when the size of the material removed decreases, the probability of encountering a stress-reducing defect decreases. Since the shear stress and strain in metal cutting is unusually high, discontinuous microcracks usually form on the metal-cutting shear plane. If the material being cut is very brittle, or the compressive stress on the shear plane is relatively low, microcracks will grow into gross cracks and give rise to discontinuous chip formation. When discontinuous microcracks form on the shear plane they weld and reform as strain proceeds, thus joining the transport of dislocations in accounting for the total slip of the shear plane. In the presence of a contaminant, the rewelding of microcracks decreases, resulting in decrease in the cutting force required for chip formation. Owing to the complexity of elastic-plastic deformation at mico & nanometer scale, the worldwide convinced precision materials removal theory is not built up until now.

Presently, there are two basic approaches on analysis of metal cutting process, namely, the theoretical method and the experimental method. As the complexity associate with the precision machining process, which involve high strains, strain rates, size effects and temperature, various simplifications and idealizations are necessary and therefore important machining features such as the strain hardening, strain rate sensitivity, temperature dependence, chip formation and the chip-tool interface behaviors are not fully accounted for by the analytical methods. Experimental studies on precision machining are expensive and time consuming. Moreover, their results are valid only for the experimental conditions used and depend greatly on the accuracy of calibration of the experimental equipment and apparatus used. Advanced numerical techniques such as Finite Element Method is a potential alternative for solving precision machining problems.

Finite Element Method (FEM) which is originated from continuum mechanics, has already been justified as successful method in analyzing complicated engineering problem. There are many advantages of using FEM to investigate machining: multi-physical machining variables output can be acquired (cutting force, chip geometry, stress and temperature distributions), improving precision and the efficiency comparing with Try-Out-Method and so on. In the last three decades, FEM has been progressively applied to metal cutting simulations. Starting with two-dimension simulations of the orthogonal cutting more than two decades ago, researches progressed to three-dimensional FEM models of the oblique cutting, which is capable of simulating metal cutting processes such as turning and milling. Increased computation power and the development of robust calculation algorithms (thus widely availability of FEM programs) are two major contributors to this progress. Unfortunately, this progress was not accompanied by new developments in precision machining theory so the age-old problems such as the chip formation mechanism and tribology of the contact surfaces are not modeled properly. Further, even at a moderate cutting speed, the strain rates are quite high, almost of the order of 10^4 per second and the temperature rise is also quite large. As a result, the visco-plasticity and temperature-softening effects become more important compared to strain-hardening. Therefore, the material properties associated with these two effects should be known for a range of strain rates and temperatures occurring in typical machining processes. Additionally, to incorporate the temperature rise in the analysis, one needs to solve the heat transfer equation governing the temperature field in conjunction with the usual three equations governing the deformation field. For plastic deformation, these equations are coupled, and hence difficult to solve.

In material removal processes at the precision scale, the undeformed chip thickness can be on the order of a few microns or less, and can approach the nanoscale in some cases. At these length scales, the surface, subsurface, and edge condition of machined features and the fundamental mechanism for chip formation are much more intimately affected by the material properties and microstructure of the workpiece material, such as ductile/brittle behavior, crystallographic orientation of the material at the tool/chip interface, and micro-topographical features such as voids, secondary phases, and interstitial particulates. Characterizing the surface, subsurface, and edge condition of machined features at the precision scale in the FEM analysis are of increasing importance for understanding, and controlling the manufacturing process. There are still many challenges in the investigation of precision machining by means of FEM. As mentioned above, this chapter will give some key factors on numerical modeling of precision machining and current advancements.

1. FINITE ELEMENT METHOD

Since the actual problem is replaced by a simpler one in finding the solution, people will be able to find only an approximate solution rather than the exact solution. The existing mathematical tools will not be sufficient to find the exact solution of most of the practical problems. Finite element method (FEM) [1] is preferred as there are not any other methods for the approximate solution of a given problem such as materials removal. The basic idea of finite element method is the division of a continuum into several simple subdomains called finite element, namely, dividing the continuum into assemble of elements with definite amounts and definite sizes for which the systematic approximate solution is constructed by applying the variational or weighted residual methods. In effect, FEM reduces the problem to that of a finite number of unknowns by dividing the domain into elements and by expressing the unknown field variable in terms of the assumed approximating functions within each element. These functions (also called interpolation functions) are defined in terms of the values of the field variables at specific points, referred to as nodes. Nodes are usually located along the element boundaries, and they connect adjacent elements at the same time some physical quantities such as displacement and force are transferred by these points. Since the actual variation of the field variables (*e.g.*, displacement, stress, temperature, pressure or velocity) inside the continuum is not known, we assume that the variation of the field variable inside a finite element can be approximated by a simple function. These approximating functions (interpolation functions) are defined in terms of values of the field variables at the nodes. When the field equations (such as equilibrium equations) for the whole continuum are written, the new unknowns will be the nodal values of the field variable. By solving the field equations, which are generally in the form of matrix equations, the nodal values of field variable will be known. Once these are known, the approximating functions define the field variable throughout the assemblage of elements.

The solution of a general continuum problem by the finite element method always follows an orderly step-by step process. With reference to general problems, the step-by step procedure can be stated as follows:

1. Discretization of the structure

    The first step in the finite element method is to divide the structure or solution region into subdivisions or elements. Hence, the structure is to be modeled with suitable finite elements. The number, type, size, and arrangement of the elements are to be decided.

2. Selection of a proper interpolation or displacement model

    Since the displacement solution of a complex structure under any specified load conditions cannot be predicted exactly, we assume some suitable solution within an element to approximate the unknown solution. The assumed solution must be simple from a computational standpoint, but it should satisfy certain convergence requirements. In general, the solution or the interpolation model is taken in the form of a polynomial.

3. Derivation of element stiffness matrices and load vectors

From the assumed displacement model, the stiffness matrix $[K^{(e)}]$ and the load vector $\vec{P}^{(e)}$ of element e are to be derived by using either equilibrium conditions or a suitable variational principle.

(i) The element strain derived from geometry equation:

$$\{\varepsilon\} = [B]\{\sigma\}^e \tag{3.1}$$

here $\{\varepsilon\}$ is strain vector of any point in the element, $[B]$ is element strain matrix, $\{\sigma\}^e$ is stress vector of any point in the element.

(ii) The element stress derived from constitutive equation

$$\{\sigma\} = [D][B]\{\delta\}^e \tag{3.2}$$

$[D]$ is materials elastic matrix.

(iii) Construct of relationship between node force and node displacement based upon variation

$$\{F\}^e = [k]^e \{\delta\}^e \tag{3.3}$$

here $\{F\}^e$ is element force vector, $[k]^e$ is the element stiffness matrix which can be acquired as following:

$$[k]^e = \iiint [B]^T [D][B] dxdydz \tag{3.4}$$

the integral domain is the volume of the element.

④ Assemblage of element equations to obtain the overall equilibrium equations

Since the structure is composed of several finite elements, the individual element stiffness matrices and load vectors are to be assembled in a suitable manner and the overall equilibrium equations have to be formulated as

$$[K^\alpha]\vec{\Phi}^\alpha = \vec{P}^\alpha \qquad\qquad (3.5)$$

where $[K^\alpha]$ is the assembled stiffness matrix, $\vec{\Phi}^\alpha$ is the vector of nodal displacements, and $\vec{P}^\alpha$ is the vector of nodal forces for the complete structure.

⑤ Solution for the unknown nodal displacements

The overall equilibrium equations have to be modified to account for the boundary conditions of the problem. After the incorporation of the boundary conditions, the equilibrium equations can be expressed as

$$[K]\vec{\Phi} = \vec{P} \qquad\qquad (3.6)$$

For linear problems, the vector $\vec{\Phi}$ can be solved very easily. However, for nonlinear problems, the solution has to be obtained in a sequence of steps, with each step involving the modification of the stiffness matrix $[K]$ and/or the load vector $\vec{P}$.

⑥ Computation of element strains and stresses

From the known nodal displacements $\vec{\Phi}$, if required, the element strains and stresses can be computed by using the necessary equations of solid or structural mechanics.

The terminology used in the previous six steps has to be modified if we want to extend the concept to other fields. For example, we have to use the term continuum or domain in place of structure, field variable in place of displacement, characteristic matrix in place of stiffness matrix, and element resultants in place of element strains.

2. PLASTIC DEFORMATION

The use of computational techniques is increasing day by day in the manufacturing sector. Process modeling and optimization with the help of computers can reduce expensive and time consuming experiments for manufacturing good quality products. Metal machining is a prominent manufacturing process which involves large deformation of elastic-plastic materials due to applied loads. In machining, the material is deformed till fracture, in order to remove material in the form of chips. To understand the physics of metal machining processes, one needs to understand the kinematics of large deformation (dependence of deformation and its rate on displacement) as well as the constitutive behavior of elastic-plastic materials (dependence of internal forces on deformation and its rate). Once the physics is understood, this phenomenon needs to be converted to mathematical relations in the form of differential equations. The interaction of the workpiece with the tools and other surroundings also needs to be expressed in a mathematical form (known as the boundary and initial conditions).

The plastic deformation theory [2] is constructed upon strain-stress experimental results and can be applicable for study materials plastic deformation behavior. Different plastic theory is selected for large deformation machining process because of different materials model. The infinitesimal deformation theory is preferable for rigid-plastic materials model because it neglecting the elastic deformation while the finite deformation theory is used for study the elastic-plastic materials. The following sections give a brief introduction of above theory.

2.1. Infinitesimal Deformation Theory

2.1.1. Stress, Strain and Strain Rate

Stress Tensor

The basic quantities that may be used to describe the mechanics of deformation when materials deforms from one configuration to another under external load are the stress, strain and strain-rate. The stress, strain and strain rate in the case of infinitesimal deformation conditions can be expressed as follows:

$$\sigma = \frac{P}{A} \tag{3.7}$$

$$\dot{\varepsilon} = \frac{v}{l} \tag{3.8}$$

$$d\varepsilon = \frac{dl}{l} \tag{3.9}$$

here v is velocity while stress defined in (3.7) is named as *Cauthy* stress. The materials total deformation can be acquired by integrate of infinitesimal strain:

$$\varepsilon = \int_0^l \frac{dl}{l} = \ln(\frac{l}{l_0}) \tag{3.10}$$

The stress status of any point in the materials should be expressed by stress tensor which is named as *Cauthy* stress tensor:

$$[\sigma_{ij}] = \begin{bmatrix} \sigma_{11} & \sigma_{21} & \sigma_{31} \\ \sigma_{12} & \sigma_{22} & \sigma_{32} \\ \sigma_{13} & \sigma_{23} & \sigma_{33} \end{bmatrix} = \begin{bmatrix} \sigma_{xx} & \sigma_{yx} & \sigma_{zx} \\ \sigma_{xy} & \sigma_{yy} & \sigma_{zy} \\ \sigma_{xz} & \sigma_{yz} & \sigma_{zz} \end{bmatrix} \tag{3.11}$$

here $[\sigma_{ij}]$ is a symmetric stress tensor which has six independent component.

Any point in the materials can be determined by three principal stresses component (stress invariant) which can be solved as follows:

$$\sigma^3 - I_1\sigma^2 - I_2\sigma - I_3 = 0 \tag{3.12}$$

where I_1, I_2 and I_3 are three stress invariant of σ_{ij} and can be determined as follows:

$$I_1 = \sigma_x + \sigma_y + \sigma_z = \sigma_1 + \sigma_2 + \sigma_3$$
$$I_2 = -(\sigma_x\sigma_y + \sigma_y\sigma_z + \sigma_z\sigma_x) + \tau_{xy}^2 + \tau_{yz}^2 + \tau_{zx}^2 = -(\sigma_1\sigma_2 + \sigma_2\sigma_3 + \sigma_3\sigma_1) \tag{3.13}$$
$$I_3 = \sigma_x\sigma_y\sigma_z + 2\tau_{xy}\tau_{yz}\tau_{zx} - \sigma_x\tau_{yz}^2 - \sigma_y\tau_{zx}^2 - \sigma_z\tau_{xy}^2$$

Strain Rate Tensor

The strain rate tensor ($\dot{\varepsilon}_{ij}$) is symmetric stress tensor which is expressed as follows:

$$\dot{\varepsilon}_x = \frac{\partial v_x}{\partial x}, \quad \dot{\varepsilon}_y = \frac{\partial v_y}{\partial y}, \quad \dot{\varepsilon}_z = \frac{\partial v_z}{\partial z}$$

$$\dot{\varepsilon}_{xy} = \frac{1}{2}\left(\frac{\partial v_x}{\partial y} + \frac{\partial v_y}{\partial x}\right) = \frac{\dot{\gamma}_{xy}}{2}$$

$$\dot{\varepsilon}_{yz} = \frac{1}{2}\left(\frac{\partial v_y}{\partial z} + \frac{\partial v_z}{\partial y}\right) = \frac{\dot{\gamma}_{yz}}{2}$$　　　(3.14)

$$\dot{\varepsilon}_{zx} = \frac{1}{2}\left(\frac{\partial v_z}{\partial x} + \frac{\partial v_x}{\partial z}\right) = \frac{\dot{\gamma}_{zx}}{2}$$

here v_i ($i=x,y,z$) is velocity component, $\dot{\gamma}_{ij}$ is engineering shear strain rate component.

2.1.2. Yielding Condition

The yielding condition is the limit of materials elastic deformation under external loading and can be expressed as follows:

$$f(\sigma_{ij}) = C(constant)$$　　　(3.15)

here $f(\sigma_{ij})$ is yielding stress function, the suitability of any proposed yield criterion must be verified by experiment.

The plastic yielding of isotropic materials can depend only on the magnitude of three principal stresses and not on their directions. Then any yield criterion is expressible in the form

$$f(I_1, I_2, I_3) = C$$　　　(3.16)

It is justified by the experimental result that the yielding of material is approximately unaffected by hydrostatic pressure or tension. The materials yielding depends only on the principal components $(\sigma_1', \sigma_2', \sigma_3')$ of the deviatoric stress tensor

$$\sigma'_{ij} = \sigma_{ij} - \delta_{ij}\sigma_m$$　　　(3.17)

where $\sigma_m = \dfrac{1}{3}(\sigma_1 + \sigma_2 + \sigma_3)$ is the hydrostatic component of the stress and δ_{ij} is *Kronecker* unit tensor. The principal components of the deviatoric stress tensor are not independent, since $\sigma'_1 + \sigma'_2 + \sigma'_3$ is identically zero.

The yield criterion then reduce to the form

$$f(J_2, J_3) = C \tag{3.18}$$

here

$$\begin{aligned} J_2 &= -(\sigma'_1\sigma'_2 + \sigma'_2\sigma'_3 + \sigma'_3\sigma'_1) \\ J_3 &= \sigma'_1\sigma'_2\sigma'_3 \end{aligned} \tag{3.19}$$

The Tresca and the Mises Yield Criteria

Two simple criteria have been in extensive use for the analysis of metal deformation, of which Tresca yielding criterion is expressed as follows:

$$\sigma_1 - \sigma_3 = constant \tag{3.20}$$

with $\sigma_1 \geq \sigma_2 \geq \sigma_3$.

Another criterion was proposed by Von Mises [3] and traditionally called Von Mises criterion. The criterion states that yielding occurs when J_2 reaches a critical value and can be written in the form:

$$J_2 = \frac{1}{2}\left(\sigma'^2_1 + \sigma'^2_2 + \sigma'^2_3\right) = \frac{1}{2}\sigma'_{ij}\sigma'_{ij} = k^2$$

or $\tag{3.21}$

$$\left(\sigma_1 - \sigma_2\right)^2 + \left(\sigma_2 - \sigma_3\right)^2 + \left(\sigma_3 - \sigma_1\right)^2 = 6k^2$$

where k is a parameter regulating the stress scale and depending on the material property. For most metals distortion energy criterion fits the data more closely than the shear stress criterion.

2.1.3. Plastic Potential and Flow Rule

Hooke's law is well known for describing the relationships between stresses and corresponding deformation in the elastic deformation regime. When the deformation extends to the plastic range, the stress and the plastic strain relationships are derived using the concept of plastic potential.

The ratios of the components of the plastic strain-rate $\dot{\varepsilon}_{ij}^{p}$ (infinitesimal plastic strain $d\varepsilon_{ij}^{p}$) are defined by

$$\dot{\varepsilon}_{ij}^{p} = h\frac{\partial g}{\partial \sigma_{ij}}f \tag{3.22}$$

$$d\dot{\varepsilon}_{ij}^{p} = h\frac{\partial g}{\partial \sigma_{ij}}df \tag{3.23}$$

here g and h are scalar functions of the invariants of deviatoric stress and f is the yield function while $g(\sigma_{ij})$ is called the plastic potential. Although the above equations are expressed in rate form, the relations between stress and strain are independent of time. If $g=f$ then

$$\dot{\varepsilon}_{ij}^{p} = \frac{\partial f}{\partial \sigma_{ij}}\dot{\lambda} \text{ or } d\dot{\varepsilon}_{ij}^{p} = \frac{\partial f}{\partial \sigma_{ij}}d\lambda \tag{3.24}$$

here $\dot{\lambda}$ or $d\lambda$ is positive proportionality constant which being equal to hf or hdf. Equation (3.24) is a flow rule associated with the yield function $f(\sigma_{ij})$. In the differentiation of $f(\sigma_{ij})$ with respect to shear stress components

$$2\tau_{xy}^{2} = \tau_{xy}^{2} + \tau_{yx}^{2} \tag{3.25}$$

The maximum plastic work principle follows from the flow rule and the convexity of the yield locus which expressed as follows:

$$\left(\sigma_{ij} - \sigma_{ij}^{\alpha}\right)\dot{\varepsilon}_{ij}^{p} \geq 0 \tag{3.26}$$

here σ_{ij} is a yield state of stress, $\dot{\varepsilon}_{ij}^{p}$ is the associate strain-rate and σ_{ij}^{α} is any other stress state represented by a point either on or inside the yield surface. Since $\dot{\varepsilon}_{ij}^{p}$ is parallel to the outward normal to the yield locus at the point σ_{ij}', $\left(\sigma_{ij}-\sigma_{ij}^{\alpha}\right)\dot{\varepsilon}_{ij}^{p}$ is proportional to the scalar product of the outward normal to the yield locus at the point σ_{ij}' with the chord joining σ_{ij}^{α} to σ_{ij}'. Therefore, equation (3.26) holds and the equality sign applies when $\sigma_{ij}' = \sigma_{ij}'^{*}$ or σ_{ij} and σ_{ij}^{*} differ by a uniform hydrostatic stress.

2.1.4. The Prandtl-Reuss Equation and the Levy-Mises Equation

The equation (3.24) can be expressed as follows if the yield function given by

$$f(\sigma_{ij}) = J_2 = \frac{1}{2}\sigma_{ij}'\sigma_{ij}'$$

$$\dot{\varepsilon}_{ij}^{p} = \sigma_{ij}'\dot{\lambda} \tag{3.27}$$

while $\dfrac{\partial f}{\partial \sigma_{ij}} = \dfrac{\partial f}{\partial \sigma_{kl}'}\dfrac{\partial \sigma_{kl}'}{\partial \sigma_{ij}} = \sigma_{ij}'$

nothing that the repeated subscripts k and l indicate summation with respect to these quantities.

Equation (3.27) can also be written as

$$\frac{\dot{\varepsilon}_x^{p}}{\sigma_x'} = \frac{\dot{\varepsilon}_y^{p}}{\sigma_y'} = \frac{\dot{\varepsilon}_z^{p}}{\sigma_z'} = \frac{\dot{\gamma}_{xy}^{p}}{2\tau_{xy}} = \frac{\dot{\gamma}_{yz}^{p}}{2\tau_{yz}} = \frac{\dot{\gamma}_{zx}^{p}}{2\tau_{zx}} \tag{3.28}$$

Combining the elastic strain-rate components $\dot{\varepsilon}_{ij}^{e}$ and the plastic strain-rate components $\dot{\varepsilon}_{ij}^{p}$, according to $\dot{\varepsilon} = \dot{\varepsilon}_{ij}^{e} + \dot{\varepsilon}_{ij}^{p}$, we obtain the Prandtl-Reuss equations for elastic-plastic solids, as

$$\dot{\varepsilon}_{ij} = \sigma_{ij}'\dot{\lambda} + \frac{1}{2G}\dot{\sigma}_{ij}' + \delta_{ij}\left(\frac{1-2v}{E}\right)\dot{\sigma}_m \tag{3.29}$$

here G, E and v are shear modulus, Yong's modulus and Possion's ration respectively.

For rigid-plastic materials, the assumption is that $\dot{\varepsilon}_{ij} \cong \dot{\varepsilon}_{ij}^p$, and the Levy-Mises equations are obtained by removing the superscripts p in equation (3.28). They are expressed in terms of stress components σ_{ij} by three equations of the type

$$\dot{\varepsilon}_x = \left\{ \sigma_x - \frac{1}{3}\left(\sigma_x + \sigma_y + \sigma_z\right) \right\} \dot{\lambda} = \left\{ \sigma_x - \frac{1}{2}\left(\sigma_x + \sigma_y + \sigma_z\right) \right\} \frac{2}{3}\dot{\lambda} \tag{3.30}$$

and three of the type

$$\dot{\varepsilon}_{xy} = \frac{\dot{\gamma}_{xy}}{2} = \tau_{xy}\dot{\lambda} \tag{3.31}$$

2.1.5. Strain-Hardening, Effective Stress and Effective Strain

Material's resistance to further deformation will increase (strain hardening) if it is deformed at room temperature. For mathematical formulation of strain-hardening, it is assumed that the final yield locus is the same no matter by any strain-path a given stress state is reached. The total plastic work per unit volume during a certain finite deformation is

$$W_p = \int \sigma_{ij} d\varepsilon_{ij} \tag{3.32}$$

where the integral is taken over the actual strain-path.

The hypothesis that the radius of the yield locus of the distortion energy criterion is a function only of W_p may be written as

$$\bar{\sigma} = \sqrt{\frac{1}{2}\left\{ \left(\sigma_1 - \sigma_2\right)^2 + \left(\sigma_2 - \sigma_3\right)^2 + \left(\sigma_3 - \sigma_1\right)^2 \right\}^{\frac{1}{2}}}$$

$$= \sqrt{\frac{3}{2}}\left(\sigma'_{ij}\sigma'_{ij}\right)^{\frac{1}{2}} = F(W_p) \tag{3.33}$$

Here $\bar{\sigma}$ is written for Y or $\sqrt{3}k$ and known as the flow stress, the effective stress, generalized stress or equivalent stress.

Another hypothesis for strain-hardening relates $\bar{\sigma}$ to a certain measure of the total plastic deformation. A quantity $d\bar{\varepsilon}$ known as the effective, generalized or equivalent infinitesimal plastic strain is defined according to

$$dW_p = \sigma_{ij}d\varepsilon_{ij} = \bar{\sigma}d\bar{\varepsilon} \tag{3.34}$$

The effective strain $\bar{\varepsilon} = \int d\bar{\varepsilon}$ which integrate over the strain path provides a measure of the plastic distortion. The strain-hardening characteristics can be formulated by

$$\bar{\sigma} = H\left(\int d\varepsilon\right) = H\left(\bar{\varepsilon}\right) \tag{3.35}$$

here H is a certain function depending on the metal concerned.

It should be noted that an explicit expression for the effective strain can be obtained when the principal axes of successive strain-increments do not rotate relative to the element, and further, when the components of any strain-increments bear constant ratios to one another. It should also be noted that although quantity $\int d\varepsilon_i$ can always be formed, even the principal axes rotate, they cannot generally be evaluated explicitly, nor do they possess geometrical significance. With the measurements of $\bar{\sigma}$ and $d\bar{\varepsilon}$ (and therefore $\dot{\bar{\varepsilon}}$, effective strain-rate), the proportionality factor $\dot{\lambda}$ in the Levy-Mises equations can be expressed by

$$\dot{\lambda} = \frac{3\dot{\bar{\varepsilon}}}{2\bar{\sigma}} \tag{3.36}$$

2.1.6. Extremum Principles

Similar with extremum principles in the theory of elasticity, two related extremum principles pertaining to a consistent state are proven for a rigid-plastic material that undergoes plastic deformation under prescribed surface traction over the surface S_F and the prescribed velocity over the surface S_u. A complete solution to this problem consists of an equilibrium stress field and an associated velocity field satisfying the boundary conditions, and the stress field satisfies the yield criterion where the deformation occurs and does not violate the yield criterion in the rigid regions. The uniqueness theorem proves that the stress field of the complete solution is uniquely determined in the deforming region.

The First Extremum Principle

Let the σ_{ij}^* be an equilibrium stress field satisfying the stress boundary conditions on S_F and nowhere violating the yield criterion. The principle of work-rate gives

$$\int_V \left(\sigma_{ij} - \sigma_{ij}^* \right) \dot{\varepsilon}_{ij} dV = \int_S \left(F_i - F_i^* \right) u_i dS \tag{3.37}$$

Then

$$\int_S \left(F_i - F_i^* \right) u_i dS \geq 0 \tag{3.38}$$

according to the maximum plastic work principle. Since $F_i - F_i^*$ on S_F, where $S = S_F + S_u$,

$$\int_{S_u} F_i u_i dS \geq \int_{S_u} F_i^* u_i dS \tag{3.39}$$

equation (3.39) is the first extremum principle and it states that a lower bound to the rate of work of the rate of the actual surface traction can be obtained from a statically admissible stress field.

In the quasi-static flow of a rigid plastic solid, velocity discontinuity is permissible. When the velocity field u_i contains surfaces of discontinuity, the principle of virtual work-rate gives

$$\int_V \left(\sigma_{ij} - \sigma_{ij}^* \right) \dot{\varepsilon}_{ij} dV + \int_{S_D} \left(k - \tau^* \right) | \Delta u | dS = \int_S \left(F_i - F_i^* \right) u_i dS \tag{3.40}$$

here S_D is a surface of velocity discontinuity, τ^* is the shear stress component of σ_{ij}^*, and k is the shear yield stress along the surface of S_D, where $| \Delta u |$ is the amount of tangential velocity discontinuity across S_D.

It should be mentioned that a surface of stress discontinuity in an equilibrium state of stress is also permissible, but that its presence does not affect the relationships described above. Further, note that a surface of stress discontinuity cannot be the surface of velocity discontinuity in the associated stress and velocity fields, because physical interpretations are the limits of the elastic region and large plastic deformation regions for stress and velocity discontinuities respectively.

The Second Extremum Principle

In the second extremum principle, let u_i^* be a velocity field that satisfies the incompressibility and the velocity boundary conditions on S_u. Such a velocity field is said to be kinematically admissible. The principle of virtual work-rate gives

$$\int_V \sigma_{ij} \dot{\varepsilon}_{ij}^* dV + \int_{S_D^*} \tau \, | \Delta u^* | \, dS = \int_S F_i u_i^* dS \tag{3.41}$$

here τ is the shear stress component of the stress σ_{ij} of the complete solution along the velocity discontinuity surface S_D^* with the amount of tangential discontinuity $| \Delta u^* |$ in the velocity field u_i^*. If u_i^* is the actual solution, then $\tau = k$, thus

$$k \, | \Delta u^* | \geq \tau \, | \Delta u^* | \tag{3.42}$$

and also

$$\sigma_{ij}^* \dot{\varepsilon}_{ij}^* \geq \sigma_{ij} \dot{\varepsilon}_{ij}^* \tag{3.43}$$

from the maximum plastic work principle, where the σ_{ij}^* satisfies the yield criterion and is associated with $\dot{\varepsilon}_{ij}^*$. Therefore, we have

$$\int_V \sigma_{ij}^* \dot{\varepsilon}_{ij}^* dV + \int_{S_D^*} k \, | \Delta u^* | \, dS \geq \int_S F_i u_i^* dS \tag{3.44}$$

The surface $S = S_F + S_u$ and on S_u, $u_i^* = u_i$, consequently

$$\int_V \sigma_{ij}^* \dot{\varepsilon}_{ij}^* dV + \int_{S_D^*} k \,|\, \Delta u^* \,|\, dS - \int_{S_F} F_i u_i^* dS \geq \int_{S_u} F_i u_i dS \tag{3.45}$$

The above equation is the second extremum principle and it shows an upper bound to the rate of work of the actual surface traction on S_u which can be obtained from a kinematically admissible velocity field.

2.2. Viscoplasticity

The mathematical theory of plasticity adequately describes the time independent aspect of the behavior of materials but is inadequate for analysis of time-dependent behavior. An approach to achieving a satisfactory formulation for time-dependent behavior has been to generalize plasticity to cases within the strain-rate-sensitive range. One such generalization has been provided by the theory of viscoplasticity. In this section, we describe an approach to the construction of of equations for the rigid viscoplastic materials. Then the extremum principle is presented as the basis for the finite element formulation of viscoplastic flow analysis.

2.2.1. Constitutive Equations

Consider a rigid material and apply the infinitesimal theory for each incremental deformation which can be expressed as follows:

$$F(\sigma_{ij}) = \frac{\sqrt{\frac{1}{2}\{\sigma_{ij}' \sigma_{ij}'\}^{1/2}}}{k} - 1 \tag{3.46}$$

here k is the static yield stress in shear process. Consider $F(\sigma_{ij})$ as the function similar to the plastic potential in the theory of plasticity, the constitutive equation is expressed as

$$\dot{\varepsilon}_{ij} = \gamma'' \langle \phi(F) \rangle \frac{\partial F}{\partial \sigma_{ij}} \tag{3.47}$$

where γ'' is a viscosity constant of material, and $\langle \phi(F) \rangle$ is a function of F such that

$$\langle \phi(F) \rangle = \begin{cases} 0, & F \leq 0 \\ \phi(F), & F > 0 \end{cases} \tag{3.48}$$

Then

$$\dot{\varepsilon}_{ij} = \gamma' < \phi\left(\frac{\overline{\sigma}}{Y} - 1\right) > \frac{\sigma'_{ij}}{\overline{\sigma}} \tag{3.49}$$

here $\overline{\sigma} = \sqrt{2/3}(\sigma'_{ij}\sigma'_{ij})^{1/2}$ so that $\overline{\sigma}$ is identical to the yield stress in uniaxial tension, and $Y = \sqrt{3}k$ is the static yield stress in tension.

Squaring both sides of equation (3.49),

$$\dot{\varepsilon}_{ij}\dot{\varepsilon}_{ij} = \gamma'^2\left(\left\langle\phi\left(\frac{\overline{\sigma}}{Y} - 1\right)\right\rangle\right)^2 \frac{\sigma'_{ij}\sigma'_{ij}}{\overline{\sigma}^2} \tag{3.50}$$

and using $\dot{\overline{\varepsilon}} = \sqrt{2/3}(\dot{\varepsilon}_{ij}\dot{\varepsilon}_{ij})^{1/2}$, we get

$$\dot{\overline{\varepsilon}} = \frac{2}{3}\gamma'\left\langle\phi\left(\frac{\overline{\sigma}}{Y} - 1\right)\right\rangle \tag{3.51}$$

and the constitutive equation becomes

$$\dot{\varepsilon}_{ij} = \frac{3}{2}\frac{\dot{\overline{\varepsilon}}}{\overline{\sigma}}\sigma'_{ij} \tag{3.52}$$

with $\overline{\sigma} = \sqrt{2/3}\{\sigma'_{ij}\sigma'_{ij}\}^{1/2}$, $\dot{\overline{\varepsilon}} = \sqrt{2/3}\{\dot{\varepsilon}_{ij}\dot{\varepsilon}_{ij}\}^{1/2}$

Equation (3.52) is formally identical to the Levy-Mises equations. The effective stress $\overline{\sigma}$ in equation (3.52) depends on the strain-rate-dependent function ϕ which is to be determined by the properties of the materials under consideration.

2.2.2. Extremum Principle

Hill [4] derived the second extremum principle associated with the deformation process of the rigid-viscoplastic materials. Among all the possible constitutive equations, the work function $E\left(\dot{\varepsilon}_{ij}\right)$ is assumed to be existed:

$$\sigma'_{ij} = \frac{\partial E}{\partial \dot{\varepsilon}_{ij}} \tag{3.53}$$

here E is convex. The existence of work function $E(\dot{\varepsilon}_{ij})$ is ensured if σ'_{ij} is single-valued function of $\dot{\varepsilon}_{ij}$ which satisfied $\partial \sigma'_{ij} / \partial \dot{\varepsilon}_{kl} = \partial \sigma'_{kl} / \partial \dot{\varepsilon}_{ij}$. Also E is a convex if

$$E(\dot{\varepsilon}^*_{ij}) - E(\dot{\varepsilon}_{ij}) \geq (\dot{\varepsilon}^*_{ij} - \dot{\varepsilon}_{ij}) \frac{\partial E}{\partial \dot{\varepsilon}_{ij}} \tag{3.54}$$

here $\dot{\varepsilon}^*_{ij}$ is a strain-rate field derived from an admissible velocity field u^*_i. With this restriction, the following relationship can be derived

$$\int_V E(\dot{\varepsilon}^*_{ij}) dV - \int_{S_F} F_i u^*_i dS \geq \int_V E(\dot{\varepsilon}_{ij}) dV - \int_{S_F} F_i u_i dS \tag{3.55}$$

here $\dot{\varepsilon}_{ij}$ and u_i are actual quantities and the starred quantities are kinematically admissible ones. It can be shown that the solution is uniquely determined at points of the body where E is strictly convex, but not necessarily in all respects elsewhere.

3. THERMAL EFFECTS AND THERMAL-MECHANICAL COUPLING

Materials removal process in MNT involves creeping flow and high temperatures. The work material properties such as viscosity, specific heat and thermal conductivity are closely related with temperature thus it ought to be an essential consideration in the numerical solutions. The power consumed in materials removal is largely converted into heat near the cutting edge of the tool, and many of the economic and technical problems of machining are caused directly or indirectly by this heating action. The cost of machining is very strongly dependent on the rate of metal removal, and costs may be reduced by increasing the cutting speed and/or the feed rate, but there are limits to the speed and feed above which the life of the tool is shortened excessively. This may not be a major constraint when machining aluminum and magnesium and certain of their alloys, in the cutting of which other problems, such as the ability to handle large quantities of fast moving chips, may limit the rate of metal removal. The bulk of cutting, however, is carried out on steel and cast iron, and it is in the cutting of these, together with the nickel-based alloys, that the most serious technical and economic problems occur. With these higher melting point metals and alloys, the tools are heated to high temperatures as materials removal rate increases and, above certain critical speed, the tools tend to collapse after a very short cutting time under the influence of stress and temperature. That heat plays an important part in machining was clearly recognized by 1982 when F. W. Taylor, in his paper "On the art of cutting metals" [5], surveyed the steps which had led to the development of the new high speed steels. These, by their ability to cut steel and iron with the tool running at a much higher temperature, raised the permissible rates of metal removal by a factor of four. The limitations imposed by cutting temperatures have been the spur to the tool materials development of the last 90 years. Problems remain however, and even with present day tool materials, cutting speeds may be limited to 30 m/min (100 ft/min) or less when cutting certain creep resistant alloys.

It is, therefore, important to understand the factors which influence the generation of heat, the flow of heat, and the temperature distribution in the tool and work material near the tool edge. The importance of the heat distribution, not just the amount of heat, was clear to J. T. Nicolson [6] who studied metal cutting in Manchester, England around the turn of the century. In "The Engineer" (1904) he stated "There is little doubt that when the laws of variation of the temperature of the shaving and tool with different cutting angles, sizes and shapes of cut, and of the rate of abrasion. are definitely determined, it will be possible to indicate how a tool should be ground in order to meet with the best efficiency. and the various conditions to be found in practice." However, determination of temperatures and temperature distribution in the vitally important region near the cutting edge is technically difficult, and progress has been slow in the more than 100 years since the problem was clearly stated. Recent research is clarifying some of the principles, but the work done so far is only the beginning of the fundamental survey that is required.

The heat energy is the energy transmitted from high temperature region to low temperature region realized by basic particles such as atom, molecular and free electron. Presently, there are three kinds of heat transmitting method, conduction, convection and radiation. In the case of precision machining, the materials plastic deformation and the friction at the tool-workpiece interface are the two important heat sources. The non-uniform distribution of temperature field at the tool tip is the result of heat conduction while the temperature reducing of local cutting area is realized by heat convection.

3.1. First Law of Thermodynamics

A system can interact with its surroundings and may transfer energy to the surrounding medium. This energy may be in the form of heat or work. Denoting the first variation of a function in terms of its variables by δ, the first law of thermodynamics states that the first variation of the heat absorbed by a system Q is equal to the increase of the differential of internal energy change U minus the first variation of the work done on the system W, all referred to unit volume, *i.e.*,

$$\delta Q = dU - \delta W \tag{3.56}$$

We assume the mathematical meaning of δ-operation as path dependent, with the energy interactions between two end states depending upon the end states as well as the path of variations. On the other hand, d-operation in mathematical sense is the variation of the function in terms of all the variables involved in the function, including the time. Note that δ-operation does not apply to the time domain, while d-operation does.

For an adiabatic system where the transferred heat through the boundaries of the system is zero, the first law of thermodynamics reduces to the conservation of mechanical energy

$$dU - \delta W = 0 \tag{3.57}$$

For a system which completes a cyclic process, the first law reduces to

$$\delta Q + \delta W = 0 \tag{3.58}$$

that is, the sum of the first variation of the heat absorbed by a system and the first variation of the work done on the system is zero. The net change of internal energy is zero, because according to the definition of a cyclic process the initial and final state variables are identical.

Equation (3.56) is a statement of conservation of energy and, therefore, it is a principal law of mechanics. The total internal energy per unit volume U is related to several other types of energies. The most conventional form is to relate it to the kinetic energy K, the gravitational energy G, and the internal energy E, all per unit volume

$$U = K + E + G \tag{3.59}$$

This concept may be extended to a system on a macroscopic or microscopic scale.

3.2. Second Law of Thermodynamics

The first law of thermodynamics as stated by equation (3.56) to (3.59) does not show the direction of the process. This law is quantitative rather than directional. To fully define the process of energy transfer, a law defining the direction of energy transfer must be stated. The second law of thermodynamics defines the direction of the process. Consider a system under a thermodynamic process. Two distinct processes are considered, reversible and irreversible. When a system can reach its initial state from the final state through the same path of intermediate states which were passed from the initial state to the final state, the process is called irreversible. Otherwise, the process is called irreversible. According to the second law of thermodynamics, when a system completes a cycle the following inequality holds

$$\oint \frac{\delta Q}{T} \leq 0 \tag{3.60}$$

where T is the absolute temperature. This is called *Clausius* inequality. The equality sign refers to the reversible process. The important conclusion of this inequality is the direction of a thermodynamic process. An irreversible process of a system along a closed path takes place in such a way that the quantity presented on the left side of equation (3.60) is always negative while for reversible process

$$\oint \frac{\delta Q_{rev}}{T} = 0 \tag{3.61}$$

The concept of entropy as a thermodynamic property which was introduced by *Clausius* directly follows from equation (3.61). The entropy change of a reversible cycle is

$$dS = \frac{\delta Q_{rev}}{T} \tag{3.62}$$

where S is a thermodynamic property. The entropy change of a system may be observed along a closed cycle. Consider a system which transfers from state (1) to state (2) through an arbitrary process while the return path from state (2) to state (1) takes place as a reversible process from equation (3.61)

$$\int_1^2 \frac{\delta Q}{T} + \int_2^1 \frac{\delta Q_{rev}}{T} \leq 0 \tag{3.63}$$

From equation (3.62)

$$\int_2^1 \frac{\delta Q_{rev}}{T} = \int_2^1 dS = S_1 - S_2 \tag{3.64}$$

The term $(S_1 - S_2)$ is called the entropy change and it is a thermodynamic property, substituting it into the equation (3.63) and we get

$$\int_1^2 \frac{\delta Q}{T} \leq S_2 - S_1 \tag{3.65}$$

The left-hand side is the entropy transfer from state (1) to state (2).

Equation (3.65) states that the entropy transfer of a thermodynamic process can never exceed its entropy change. Equation (3.63) with the aid of equation (3.65) can be stated as follows

$$S_{gen} = \left(S_2 - S_1\right) - \int_1^2 \frac{\delta Q}{T} \geq 0 \tag{3.66}$$

where S_{gen} is the generated entropy in the defined closed cycle. This inequality states that in an arbitrary closed cycle, the thermodynamic process takes place in such a way that the total entropy generation is never negative. Equation (3.66) can be modified for a thermodynamic differential path. The total entropy change dS of a differential path of a system is divided into two parts: dS_e is the part of entropy change of the system with environment, and dS_i is the entropy change inside the system

$$dS = dS_e + dS_i \qquad\qquad (3.67)$$

The entropy exchange of the system with environment is related to the absorbed heat by

$$dS_e = \frac{\delta Q}{T} \qquad\qquad (3.68)$$

For the reversible process dS_i is zero, while for the irreversible process dS_i is always positive,

$dS_i=0$ for reversible process

$dS_i>0$ for irreversible process

For any thermodynamic state, two independent properties will define the system. The temperature and the entropy, called the fundamental properties, can fully define the thermodynamic state. While the temperature is an intensive property of the system, the entropy is an attribute of the matter and thus is an extensive property.

3.3. Temperature Field

The temperature is a physical parameter which is used to describe the cold or hot status of materials. But temperature itself cannot express the cold or hot status at every position in the bulk materials. In the precision machining, the *work* which is used to overcome materials plastic deformation, the friction between tool rake face and the chip and the friction between tool face and the machined surface are all transformed into cutting heat. The cutting heat is transferred from the heat sources to adjacent chip, tool and workpiece and raise the temperature of bulk materials. The temperature field is the distribution of temperature at every site in the definite space which can be expressed as the function of the time and space as follows [6]:

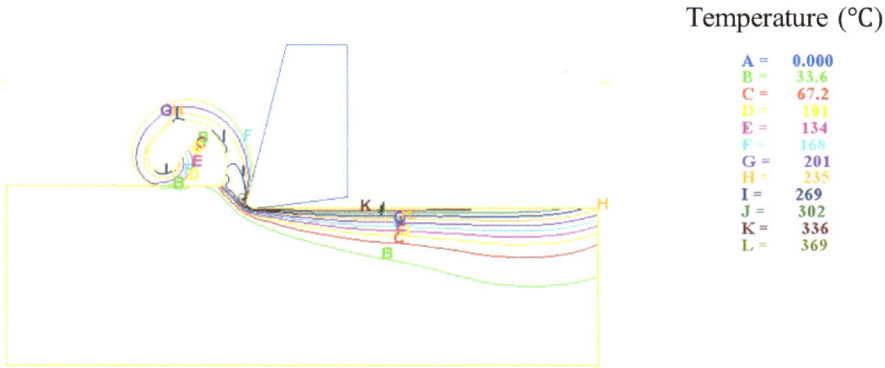

Temperature (°C)

A =	0.000
B =	33.6
C =	67.2
D =	101
E =	134
F =	168
G =	201
H =	235
I =	269
J =	302
K =	336
L =	369

Figure 1: Temperature field of micromachining.

$$\theta = f(x, y, z, t) \tag{3.69}$$

The temperature field is a scalar field because the temperature is a scalar physical quantity. Generally speaking, the temperature field in the uniform body is a continuum field while this continuum can be terminated at the inner crack or interface of two different elements. The sets of points with the same temperature form three dimensional plane and called isothermal surface. There is no heat flow along the tangent plane of isothermal surface as the heat can only be transmitted from site with high temperature to the site with low temperature (Fig. **1**).

3.3.1. Differential Equation of Heat Conduction

The heat conduction is the result of non-uniformly distribution of temperature field which can be mathematically expressed as follows:

$$q = q_x + q_y + q_z = \left(-\lambda_x \frac{\partial \theta}{\partial x} \right) + \left(-\lambda_y \frac{\partial \theta}{\partial y} \right) + \left(-\lambda_z \frac{\partial \theta}{\partial z} \right) \tag{3.70}$$

here $\lambda_x = \lambda_y = \lambda_z$ if the solid is a isotropic material, and the above equation can be expressed as

$$q = -\lambda \, grad\theta \tag{3.71}$$

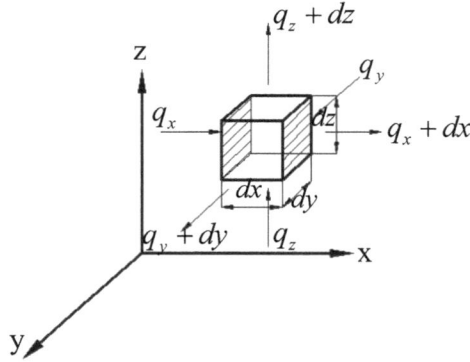

Figure 2: Micro-Unit of heat conduction

So the temperature distribution in the solids can be acquired by solving the common differential equation. Fig. **2** shows the micro-unit divided from the materials interior which is used for heat conduction analysis. The volume (*dv*) of this unit can be acquired as *dv=dxdydz*, so the input heat from left plane (with shadow) during *dt* can be expressed as follows:

$$Q_x dt = q_x dydzdt \tag{3.72}$$

While the output heat from the right plane (with shadow) during *dt* can be expressed as follows:

$$Q_{x+dx} dt = q_{x+dx} dydzdt \tag{3.73}$$

while $q_{x+dx} = q_x + \dfrac{\partial q_x}{\partial x} dx$

The net heat transmitted into the micro-unit is

$$dQ_x dt = Q_x dt - Q_{x+dx} dt = -\frac{\partial q_x}{\partial x} dxdydzdt \tag{3.74}$$

For the same reason we can get

$$dQ_y dt = -\frac{\partial q_y}{\partial x} dxdydzdt \tag{3.75}$$

$$dQ_z dt = -\frac{\partial q_z}{\partial x} dxdydzdt \tag{3.76}$$

Then the total heat transmitted into the materials during dt is

$$\left(dQ_x + dQ_y + dQ_z\right)dt = -\left(\frac{\partial q_x}{\partial x} + \frac{\partial q_y}{\partial y} + \frac{\partial q_z}{\partial z}\right)dvdt \tag{3.77}$$

If there are internal heat source, then the heat generated within dt will be $q_v dvdt$.

Based on *total differential theory*, the temperature variation of the materials can be expressed as follows

$$d\theta = \frac{\partial \theta}{\partial t}dt + \frac{\partial \theta}{\partial x}dx + \frac{\partial \theta}{\partial y}dy + \frac{\partial \theta}{\partial z}dz \tag{3.78}$$

or

$$\frac{d\theta}{dt} = \frac{\partial \theta}{\partial t} + \frac{\partial \theta}{\partial x}\frac{dx}{dt} + \frac{\partial \theta}{\partial y}\frac{dy}{dt} + \frac{\partial \theta}{\partial z}\frac{dz}{dt} \tag{3.79}$$

where $\theta = f(x, y, z, t)$ is the function of local temperature in the materials, $\frac{dx}{dt}$, $\frac{dy}{dt}$, $\frac{dz}{dt}$ hold the meaning of velocity, so the equation (3.79) can be expressed as

$$\frac{D\theta}{Dt} = \frac{\partial \theta}{\partial t} + \left(W_x \frac{\partial \theta}{\partial x} + W_y \frac{\partial \theta}{\partial y} + W_z \frac{\partial \theta}{\partial z}\right) \tag{3.80}$$

the above equation gives the total rate of change of materials temperature while the $\frac{\partial \theta}{\partial t}$ denote the local variation and the $\left(W_x \frac{\partial \theta}{\partial x} + W_y \frac{\partial \theta}{\partial y} + W_z \frac{\partial \theta}{\partial z}\right)$ denote the convection resulted by change of position.

The total heat flow during dt can be expressed as follows

$$\rho C_p \frac{D\theta}{Dt} dt dv = -\left(\frac{\partial q_x}{\partial x} + \frac{\partial q_y}{\partial y} + \frac{\partial q_z}{\partial z} \right) dv dt + q_v dv dt$$

or

$$\rho C_p \frac{D\theta}{Dt} = -\left(\frac{\partial q_x}{\partial x} + \frac{\partial q_y}{\partial y} + \frac{\partial q_z}{\partial z} \right) + q_v \qquad (3.81)$$

The above equation can also be written

$$\rho C_p \frac{D\theta}{Dt} = -\text{div}q + q_v \qquad (3.82)$$

where $\text{div}q = \dfrac{\partial q_x}{\partial x} + \dfrac{\partial q_y}{\partial y} + \dfrac{\partial q_z}{\partial z}$ is the divergence of vector field of heat flow q.

Combined the (3.82) with Fourier heat conduction model ($q_x = -\lambda \dfrac{d\theta}{dx}$) we can get

$$\rho C_p \frac{D\theta}{Dt} = \frac{\partial}{\partial x}\left(\lambda_x \frac{\partial \theta}{\partial x} \right) + \frac{\partial}{\partial y}\left(\lambda_y \frac{\partial \theta}{\partial y} \right) + \frac{\partial}{\partial z}\left(\lambda_z \frac{\partial \theta}{\partial z} \right) \qquad (3.83)$$

This is the heat conduction differential equation.

3.3.2. Boundary Conditions

The initial condition and the boundary condition are the "regional conditions" of integration of differential equation of heat conduction. The initial condition is the regional condition on time domain which describes the temperature distribution of definite region while time coordinate is zero. The boundary condition is the regional condition on space domain which describes the temperature of the regional boundary. In reality, there are three kinds of temperature boundary conditions listed as follows:

1. The first kind of boundary condition

The temperature of the boundary is known and is the function of time and position which can be expressed as follows:

$$\theta = f_i(x, y, z, t) \text{ on surface } S_i \tag{3.84}$$

or

$$\theta = f_i(x, y, z), \ \theta = f_i(t), \ \theta = 0$$

    2.   The second kind of boundary condition

The normal derivative of the boundary surface is given, which may be the function of time or position and can be expressed as follows:

$$\frac{\partial \theta}{\partial n_i} = f_i(x, y, z, t) \text{ on surface } S_i \tag{3.85}$$

where $\dfrac{\partial \theta}{\partial n_i}$ is the derivative along normal line which is the same as the heat flow density.

    3.   The third kind of boundary condition

The linear combination of temperature and its normal derivative is given which is expressed as follows:

$$\lambda_i \frac{\partial \theta}{\partial n_i} + \alpha_i \theta = f_i(x, y, z, t) \tag{3.86}$$

The first and the second boundary condition can be derived from equation (3.86) by set λ_i or α_i to zero separately. The physical meaning of third boundary conditions is that the heat is transmitted to the environment under the *newton cooling theory* at the investigation boundary.

3.4. Solving Temperature Field Using Finite Element Method

The solving process involves construct functional according to heat conduction differential equation and boundary conditions. After that the functional of each element will be solved and thus acquire the corresponding functional value of each node (temperature).

3.4.1. The Differential Equation of Heat Conduction

The general differential equation of heat conduction can be expressed as follows

$$\frac{\partial}{\partial x}\left(\lambda_x \frac{\partial \theta}{\partial x}\right) + \frac{\partial}{\partial y}\left(\lambda_y \frac{\partial \theta}{\partial y}\right) + \frac{\partial}{\partial z}\left(\lambda_z \frac{\partial \theta}{\partial z}\right) + q_v = \rho C_p \frac{\partial \theta}{\partial t} \tag{3.87}$$

where λ_x, λ_y, λ_z are the coefficients of heat conduction along x, y, z direction separately, and the boundary condition of the equation (3.87) is

$$\left(\lambda_x \frac{\partial \theta}{\partial x}\cos\alpha + \lambda_y \frac{\partial \theta}{\partial y}\cos\beta + \lambda_z \frac{\partial \theta}{\partial z}\cos\gamma\right) + \alpha_s(\theta_s - \theta_f) + q_s = 0 \tag{3.88}$$

here $\cos\alpha$, $\cos\beta$, $\cos\gamma$ are the direction cosine of boundary surface outer normal line, q_s is the heat intensity of boundary surface per unit.

Equation (3.87) and (3.88) called Euler equation and the corresponding functional can be written as

$$I = \iiint_v \left[\lambda_x \left(\frac{\partial \theta}{\partial x}\right)^2 + \lambda_y \left(\frac{\partial \theta}{\partial y}\right)^2 + \lambda_z \left(\frac{\partial \theta}{\partial z}\right)^2 - 2\theta\left(q_v - \rho C_p \frac{\partial \theta}{\partial t}\right)\right]dv + \iint_s \left[\alpha_s \theta^2 + 2(q_s - \alpha_s \theta_f)\theta\right]ds \tag{3.89}$$

The functional of third kind boundary condition of stable thermal field without inner heat source can be written as

$$I = \iiint_v \left[\lambda_x \left(\frac{\partial \theta}{\partial x}\right)^2 + \lambda_y \left(\frac{\partial \theta}{\partial y}\right)^2 + \lambda_z \left(\frac{\partial \theta}{\partial z}\right)^2\right]dv + \iint_s \left[\alpha_s \theta^2 + 2(q_s - \alpha_s \theta_f)\theta\right]ds \tag{3.90}$$

If the coefficient of heat conduction is isotropic then the above equation can be expressed as

$$I = \iiint_v \left[\left(\frac{\partial \theta}{\partial x} \right)^2 + \left(\frac{\partial \theta}{\partial y} \right)^2 + \left(\frac{\partial \theta}{\partial z} \right)^2 \right] dv + \iint_s \left[\frac{\alpha_s}{\lambda} \theta^2 - 2 \frac{\alpha_s}{\lambda} \theta_f \theta \right] ds \qquad (3.91)$$

If we make $\sigma = \dfrac{\alpha_s}{\lambda}$, $\Gamma = \dfrac{\alpha_s}{\lambda} \theta_f$ then the (3.91) can be written as

$$I = \iiint_v \left[\left(\frac{\partial \theta}{\partial x} \right)^2 + \left(\frac{\partial \theta}{\partial y} \right)^2 + \left(\frac{\partial \theta}{\partial z} \right)^2 \right] dv + \iint_s \left[\sigma \theta^2 - 2\Gamma \theta_f \theta \right] ds \qquad (3.92)$$

The second boundary conditions can be expressed as

$$I = \iiint_v \left[\left(\frac{\partial \theta}{\partial x} \right)^2 + \left(\frac{\partial \theta}{\partial y} \right)^2 + \left(\frac{\partial \theta}{\partial z} \right)^2 \right] dv + \iint_s \frac{2}{\lambda} q_s \theta ds \qquad (3.93)$$

The first boundary conditions can be expressed as

$$I = \iiint_v \left[\left(\frac{\partial \theta}{\partial x} \right)^2 + \left(\frac{\partial \theta}{\partial y} \right)^2 + \left(\frac{\partial \theta}{\partial z} \right)^2 \right] dv \qquad (3.94)$$

3.4.2. The Variation Theorem

If we solving the extreme value of equation (3.89) then we get

$$\delta I = \delta \left\{ \iiint_v \left[\lambda_x \left(\frac{\partial \theta}{\partial x} \right)^2 + \lambda_y \left(\frac{\partial \theta}{\partial y} \right)^2 + \lambda_z \left(\frac{\partial \theta}{\partial z} \right)^2 - 2\theta \left(q_v - \rho C_p \frac{\partial \theta}{\partial t} \right) \right] dv + \iint_s \left[\alpha_s \theta^2 + 2 \left(q_s - \alpha_s \theta_f \right) \theta \right] ds \right\} = 0 \qquad (3.95)$$

here δI is the variation of functional I while the independent variable of δI is θ, the variation of θ is $\delta \theta$

$$\delta \left(\frac{\partial \theta}{\partial x} \right)^2 = 2 \left(\frac{\partial \theta}{\partial x} \right) \frac{\partial (\delta \theta)}{\partial x}$$

$$\delta \left(\frac{\partial \theta}{\partial y} \right)^2 = 2 \left(\frac{\partial \theta}{\partial y} \right) \frac{\partial (\delta \theta)}{\partial y} \qquad (3.96)$$

$$\delta \left(\frac{\partial \theta}{\partial z} \right)^2 = 2 \left(\frac{\partial \theta}{\partial z} \right) \frac{\partial (\delta \theta)}{\partial z}$$

thus we get

$$\delta I = \iiint_v \left[2\lambda_x \frac{\partial \theta}{\partial x} \frac{\partial(\delta\theta)}{\partial x} + 2\lambda_y \frac{\partial \theta}{\partial y} \frac{\partial(\delta\theta)}{\partial y} + 2\lambda_z \frac{\partial \theta}{\partial z} \frac{\partial(\delta\theta)}{\partial z} - 2\left(q_v - \rho C_p \frac{\partial \theta}{\partial t} \right) \delta\theta \right] dv \quad \textbf{(3.97)}$$
$$+ \iint_s 2\left[\alpha_s \theta + (q_s - \alpha_s \theta_f) \right] \delta\theta ds$$

as

$$\frac{\partial}{\partial x}\left(\lambda_x \frac{\partial \theta}{\partial x} \delta\theta \right) = \left[\frac{\partial}{\partial x}\left(\lambda_x \frac{\partial \theta}{\partial x} \right) \right] \delta\theta + \left(\lambda_x \frac{\partial \theta}{\partial x} \right) \frac{\partial(\delta\theta)}{\partial x} \quad \textbf{(3.98)}$$

Thus

$$\left(\lambda_x \frac{\partial \theta}{\partial x} \right) \frac{\partial(\delta\theta)}{\partial x} = \frac{\partial}{\partial x}\left(\lambda_x \frac{\partial \theta}{\partial x} \delta\theta \right) - \left[\frac{\partial}{\partial x}\left(\lambda_x \frac{\partial \theta}{\partial x} \right) \right] \delta\theta \quad \textbf{(3.99)}$$

And also the process we can get

$$\left(\lambda_y \frac{\partial \theta}{\partial y} \right) \frac{\partial(\delta\theta)}{\partial y} = \frac{\partial}{\partial y}\left(\lambda_y \frac{\partial \theta}{\partial y} \delta\theta \right) - \left[\frac{\partial}{\partial y}\left(\lambda_y \frac{\partial \theta}{\partial y} \right) \right] \delta\theta \quad \textbf{(3.100)}$$

$$\left(\lambda_z \frac{\partial \theta}{\partial z} \right) \frac{\partial(\delta\theta)}{\partial z} = \frac{\partial}{\partial x}\left(\lambda_z \frac{\partial \theta}{\partial z} \delta\theta \right) - \left[\frac{\partial}{\partial z}\left(\lambda_z \frac{\partial \theta}{\partial z} \right) \right] \delta\theta \quad \textbf{(3.101)}$$

Substitute equation (3.99) into (3.97)

$$\delta I = 2\iiint_v \left[\frac{\partial}{\partial x}\left(\lambda_x \frac{\partial \theta}{\partial x} \delta\theta \right) + \frac{\partial}{\partial y}\left(\lambda_y \frac{\partial \theta}{\partial y} \delta\theta \right) + \frac{\partial}{\partial z}\left(\lambda_z \frac{\partial \theta}{\partial z} \delta\theta \right) \right] dv$$
$$- 2\iiint_v \left[\frac{\partial}{\partial x}\left(\lambda_x \frac{\partial \theta}{\partial x} \right) + \frac{\partial}{\partial y}\left(\lambda_y \frac{\partial \theta}{\partial y} \right) + \frac{\partial}{\partial z}\left(\lambda_z \frac{\partial \theta}{\partial z} \right) \right] \delta\theta dv \quad \textbf{(3.102)}$$
$$- 2\iiint_v \left(q_v - \rho C_p \frac{\partial \theta}{\partial t} \right) \delta\theta dv + 2\iint_s \left[\alpha_s \theta + (q_s - \alpha_s \theta_f) \right] \delta\theta ds$$

According to Остроградский-Gauss equation, there exists following relationship when the volume enveloped by surface S is V and the direction cosine of the outer normal line are $\cos\alpha$, $\cos\beta$, $\cos\gamma$ separately, then

$$\iint_s \left(P\cos\alpha + Q\cos\beta + R\cos\gamma \right) ds = \iiint_v \left(\frac{\partial P}{\partial x} + \frac{\partial Q}{\partial y} + \frac{\partial R}{\partial z} \right) dv \qquad (3.103)$$

here

$$P = \lambda_x \frac{\partial\theta}{\partial x}\delta\theta, \ Q = \lambda_y \frac{\partial\theta}{\partial y}\delta\theta, \ R = \lambda_z \frac{\partial\theta}{\partial z}\delta\theta \qquad (3.104)$$

and thus

$$\iiint_v \left[\frac{\partial}{\partial x}\left(\lambda_x \frac{\partial\theta}{\partial x}\delta\theta \right) + \frac{\partial}{\partial y}\left(\lambda_y \frac{\partial\theta}{\partial y}\delta\theta \right) + \frac{\partial}{\partial z}\left(\lambda_z \frac{\partial\theta}{\partial z}\delta\theta \right) \right] dv$$
$$= \iint_s \left[\lambda_x \frac{\partial\theta}{\partial x}\cos\alpha + \lambda_y \frac{\partial\theta}{\partial y}\cos\beta + \lambda_z \frac{\partial\theta}{\partial z}\cos\gamma \right] \delta\theta ds \qquad (3.105)$$

We get variation of I as follows:

$$\delta I = -2\iiint_v \left[\frac{\partial}{\partial x}\left(\lambda_x \frac{\partial\theta}{\partial x} \right) + \frac{\partial}{\partial y}\left(\lambda_y \frac{\partial\theta}{\partial y} \right) + \frac{\partial}{\partial y}\left(\lambda_y \frac{\partial\theta}{\partial y} \right) + \left(q_v - \rho C_p \frac{\partial\theta}{\partial t} \right) \right] \delta\theta dv$$
$$+ 2\iint_s \left[\left(\lambda_x \frac{\partial\theta}{\partial x}\cos\alpha + \lambda_y \frac{\partial\theta}{\partial y}\cos\beta + \lambda_z \frac{\partial\theta}{\partial z}\cos\gamma \right) + \alpha_s \left(\theta - \theta_f \right) + q_s \right] \delta\theta ds \qquad (3.106)$$

The δI must set to zero if we are to acquire the extreme value of functional I and satisfy the following conditions:

1. The first part of equation (3.106) must equal to zero

$$\frac{\partial}{\partial x}\left(\lambda_x \frac{\partial\theta}{\partial x} \right) + \frac{\partial}{\partial y}\left(\lambda_y \frac{\partial\theta}{\partial y} \right) + \frac{\partial}{\partial y}\left(\lambda_y \frac{\partial\theta}{\partial y} \right) + q_v = \rho C_p \frac{\partial\theta}{\partial t} \qquad (3.107)$$

Equation (3.107) is the common heat conduction differential equation.

2. The second part of equation (3.106) must equal to zero

$$\left(\lambda_x \frac{\partial \theta}{\partial x} \cos \alpha + \lambda_y \frac{\partial \theta}{\partial y} \cos \beta + \lambda_z \frac{\partial \theta}{\partial z} \cos \gamma \right) + \alpha_s \left(\theta - \theta_f \right) + q_s = 0 \tag{3.108}$$

Equation (3.108) is the common boundary conditions of heat conduction.

The extreme value of functional I (temperature field) is the definite solution of equation (3.88) and (3.89). The functional of each element can be constructed after set up the conception of functional of temperature.

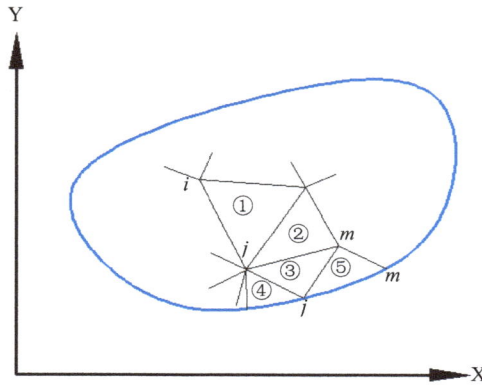

Figure 3: Triangular element division of the two-dimensional region.

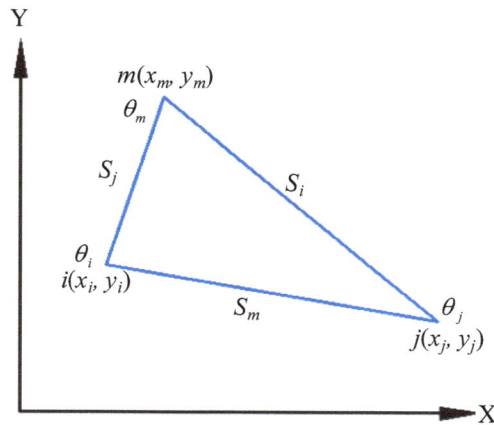

Figure 4: The temperature at the nodes of element.

3.4.3. Dividing the Temperature Field by Finite Element

Take the example of stable planar heat source and divide the interested area into triangular element as shown in Fig. (**3**). As for each element, the three vertex of triangle is arranged in the anti-clockwise order with alphabet *i*, *j*, *m* and not include the element at the boundary. The element ①, ② and ③ are called inner elements while the element ④ and ⑤ are called boundary elements and the only one edge located at the boundary and assign as *jm* which faces vertex *i*. Fig. (**4**) shows the arbitrary element of region *D* in the Fig. (**3**) while the coordinates of the three vertexes are given. The temperature for any point in the triangle can be expressed by temperature of three vertexes

$$\theta = f(\theta_i, \theta_j, \theta_m) \tag{3.109}$$

3.4.4. The Element Function and the Computation of Variation

The determination of function θ is of great important in the case of computation of variation. The trigonometric polynomial function is often selected as the key component of θ in the case of computation of classical variation in order to satisfy the total region *D*. The distribution of θ in the region *D* can be considered as linear because the region *D* has already been divided into very small part and the temperature of the element *D* can be assumed as the linear function of *x* and *y*

$$\theta = \alpha_1 + \alpha_2 x + \alpha_3 y \tag{3.110}$$

here α_1, α_2, α_3 are the constants and can be determined by the temperature of the element nodes and the coordinates of nodes

$$\begin{cases} \theta_i = \alpha_1 + \alpha_2 x_i + \alpha_3 y_i \\ \theta_j = \alpha_1 + \alpha_2 x_j + \alpha_3 y_j \\ \theta_m = \alpha_1 + \alpha_2 x_m + \alpha_3 y_m \end{cases} \tag{3.111}$$

The above equation can be written in matrix form as follows:

$$\begin{bmatrix} 1 & x_i & y_i & z_i \\ 1 & x_j & y_j & z_j \\ 1 & x_m & y_m & z_m \end{bmatrix} \begin{Bmatrix} \alpha_1 \\ \alpha_2 \\ \alpha_3 \end{Bmatrix} = \begin{Bmatrix} \theta_i \\ \theta_j \\ \theta_m \end{Bmatrix} \tag{3.112}$$

Then α_1, α_2, α_3 can be solved by inverse matrix

$$
\begin{Bmatrix} \alpha_1 \\ \alpha_2 \\ \alpha_3 \end{Bmatrix} = \begin{bmatrix} 1 & x_i & y_i & z_i \\ 1 & x_j & y_j & z_j \\ 1 & x_m & y_m & z_m \end{bmatrix}^{-1} \begin{Bmatrix} \theta_i \\ \theta_j \\ \theta_m \end{Bmatrix}
$$

$$
= \frac{1}{2\Delta_e} \begin{bmatrix} x_j y_m - x_m y_j & x_m y_i - x_i y_m & x_i y_i - x_j y_j \\ y_j - y_m & y_m - y_i & y_i - y_j \\ x_m - x_j & x_i - x_m & x_j - x_i \end{bmatrix} \begin{Bmatrix} \theta_i \\ \theta_j \\ \theta_m \end{Bmatrix}
$$

$$(3.113)$$

The equation (3.113) can be written in another form

$$
\begin{Bmatrix} \alpha_1 \\ \alpha_2 \\ \alpha_3 \end{Bmatrix} = \frac{1}{2\Delta_e} \begin{bmatrix} a_i & a_j & a_m \\ b_i & b_j & b_m \\ c_i & c_j & c_m \end{bmatrix} \begin{Bmatrix} \theta_i \\ \theta_j \\ \theta_m \end{Bmatrix}
$$

$$(3.114)$$

here

$$
\Delta_e = \frac{1}{2} \begin{vmatrix} 1 & x_i & y_i \\ 1 & x_j & y_j \\ 1 & x_m & y_m \end{vmatrix} = \frac{1}{2}\left(b_i c_j - b_j c_i \right)
$$

$$(3.115)$$

The temperature interpolation function of element e can be expressed as follows:

$$
\theta = \frac{1}{2\Delta_e} \begin{bmatrix} 1 & x & y \end{bmatrix} \begin{bmatrix} a_i & a_j & a_m \\ b_i & b_j & b_m \\ c_i & c_j & c_m \end{bmatrix} \begin{Bmatrix} \theta_i \\ \theta_j \\ \theta_m \end{Bmatrix}
$$

$$
= \frac{1}{2\Delta_e}\left[\left(a_i + b_i x + c_i y \right)\theta_i + \left(a_j + b_j x + c_j y \right)\theta_j + \left(a_m + b_m x + c_m y \right)\theta_m \right]
$$

$$(3.116)$$

$$
= \frac{1}{2\Delta_e}\left[\sum_{i,j,m} \left(a_i + b_i x + c_i y \right)\theta_i \right]
$$

4. MATERIALS REMOVAL CRITERION

Presently, two FE methods exist for analyzing the precision machining process. In the first method, it is assumed that the chip formation is continuous and the shape of the chip is known in advance. Thus, the process is analyzed as a steady-state process. This method is called Eulerian method. In this method, a chip separation criterion is not required. In the second method, the process is analyzed from the beginning to the steady state chip formation which is called Updated Lagrangian Formulation. In this method, a chip separation criterion is required to predict the chip geometry. Early applications of finite element method to the machining process were mainly Eulerian method. The main objective of many of these studies was to predict the temperature distribution and therefore, the determination of deformation and stress fields was only an intermediate step. These studies considered the machined material as rigid-plastic. But, later applications of Eulerian formulation to machining process also included viscoplastic effects. All of these applications have considered only orthogonal machining. The first finite element study of the machining process using an modified Lagrangian Formulation was made by Strenkowski and Carrol [7]. A critical value of the equivalent plastic strain was used to model the separation of a chip. Later on, several researchers used the Updated Lagrangian Formulation for analyzing two- and three-dimensional machining processes. The criterion used for chip separation has been based on controlled crack propagation or some geometrical considerations. Remeshing technique has been used to simulate the chip formation.

As the size of the material removed decreases in the precision machining, the probability of encountering a stress-reducing defect decreases. There are some new disciplines dominate the chip separation process. The metal cutting process is different from general metal processing technique as there are always accompanied with chip separation or materials removal phenomenon. The separation of chip is of utmost important about numerical simulation of precision machining. The simulation results can only be meaningful only if the reasonable chip separation criteria which can reflect materials mechanical and physical property (such as morphology of chip, force, temperature and the residual stress *etc.*) were applied in the simulation model. Besides, the criterion for chip separation should be invariant for definite materials but not change with the different working conditions. In the metal cutting process, some kinds of materials may generate continuous chip while others may generate saw-like chip thus different materials fracture criteria should be included in the finite element model.

Presently, there are two kinds of chip separation criteria, namely, the geometric criterion and the physical criterion. Materials removal (chip separation) using geometric criterion is realized through the variation of size of deformable body. On the other hand, the physical criterion is based on if some key physical parameters approached the critical value, these physical criterion includes effective plastic strain criterion, strain energy density criterion and the fracture stress criterion and so on.

4.1. Fracture Mechanics Criterion (Physical Criterion)

4.4.1. Stress Intensity Factor

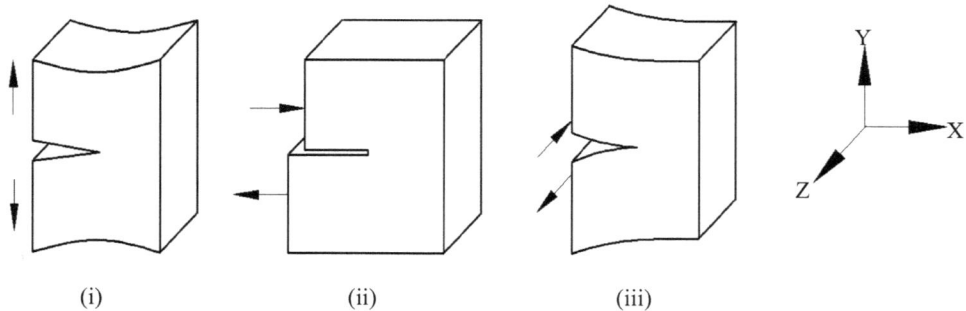

Figure 5: Three basic modes of crack extension (i) Opening mode; (ii) Sliding mode; (iii) Tearing mode.

In reality, chip separation process can be assumed as the formation and development of crack. Under what conditions and what manners can the materials be cut off is closely related with the fracture criterion [8-9]. Consider plane crack extending through the thickness of flat plane. There are three independent kinematic movements of the upper and lower crack surfaces with respect to each other. These three basic modes of deformation are illustrated in Fig. (**5**), which presents the displacements of the crack surface of a local element containing the crack front. Any deformation of the crack surface can be viewed as a superposition of these basic deformation modes, which are defined as follows:

1. Opening mode, the crack surfaces separate symmetrically with respect to the planes *xy* and *xz*

2. Sliding mode, the crack surfaces slide relative to each other symmetrically with respect to the planes *xy* and skew-symmetrically with respect to the planes *xy* and *xz*

3. Tearing mode, the crack surfaces slide relative to each other skew-symmetrically with respect to both planes *xy* and *xz*.

The stress and deformation fields associated with each of these three deformation modes will be determined in the sequel for the case of plane strain and generalized plane stress. Solid materials is defined to be in a state of plane strain parallel to the plane *xy* if

$$u=u(x,y),\ v=v(x,y),\ w=0 \tag{3.117}$$

where *u*, *v*, *w* denote the displacement components along the axis *x*, *y* and *z*. Chip separation originated from crack while the static, stable or extension of the crack are all closely related with the distribution of stress field around the crack. The study of stress field near the crack tip is of great important as this field govern the fracture process that takes place at the crack tip.

(a) Opening Mode

Infinite plate with a crack of length 2*a* subjected to equal stresses σ at infinity is give by

$$Z_I(z) = \frac{\sigma z}{\sqrt{z^2 - a^2}} \tag{3.118}$$

If we place the origin of the coordinate system at the crack tip *z=a* through the transformation

$$\varsigma = z - a \tag{3.119}$$

Then the equation (3.118) takes the form

$$Z_I = \frac{\sigma(\varsigma + a)}{\sqrt{\varsigma(\varsigma + 2a)}}$$

(3.120)

using polar coordinates, r and θ we have

$$\varsigma = re^{i\theta}$$

(3.121)

the stress near the crack tip can be derived as follows:

$$\sigma_x = \frac{K_I}{\sqrt{2\pi r}}\cos\frac{\theta}{2}\left(1 - \sin\frac{\theta}{2}\sin\frac{3\theta}{2}\right)$$

(3.122)

$$\sigma_y = \frac{K_I}{\sqrt{2\pi r}}\cos\frac{\theta}{2}\left(1 + \sin\frac{\theta}{2}\sin\frac{3\theta}{2}\right)$$

(3.123)

$$\tau_{xy} = \frac{K_I}{\sqrt{2\pi r}}\cos\frac{\theta}{2}\sin\frac{\theta}{2}\cos\frac{3\theta}{2}$$

(3.124)

$$u = \frac{K_I}{4G}\sqrt{\frac{r}{2\pi}}\left[(2\beta - 1)\cos\frac{\theta}{2} - \cos\frac{3\theta}{2}\right]$$

(3.125)

$$v = \frac{K_I}{4G}\sqrt{\frac{r}{2\pi}}\left[(2\beta + 1)\sin\frac{\theta}{2} - \sin\frac{3\theta}{2}\right]$$

(3.126)

$$w = 0$$

(3.127)

here σ_x, σ_y and τ_{xy} are the stress component, u, v and w are the displacement component, G is the shear modulus, μ is the poisson ratio, $\beta = 3 - 4\mu$. The K_I is the stress intensity factor and expresses the strength of the singular elastic stress field. As put forward by Irwin [10], equation (3.122) ~ (3.124) applies to all crack tip stress fields independently of crack/body geometry and the loading conditions. The stress intensity factor depends linearly on the applied load and is a function of crack length and the geometrical configuration of the cracked body.

$$K_I = \lim_{|\varsigma| \to 0}\sqrt{2\pi\varsigma}Z_I$$

(3.128)

Equation (3.128) can be used to determine the K_I stress intensity factor when the Z_I is known.

(b) Sliding Mode

Following the same procedure in the previous case, and recognizing the general applicability of the singular solution for all sliding mode crack problems, the following equations for stresses and displacements are obtained:

$$\sigma_x = -\frac{K_{II}}{\sqrt{2\pi r}}\sin\frac{\theta}{2}\left(2+\cos\frac{\theta}{2}\cos\frac{3\theta}{2}\right) \tag{3.129}$$

$$\sigma_y = \frac{K_{II}}{\sqrt{2\pi r}}\sin\frac{\theta}{2}\cos\frac{\theta}{2}\cos\frac{3\theta}{2} \tag{3.130}$$

$$\tau_{xy} = \frac{K_{II}}{\sqrt{2\pi r}}\cos\frac{\theta}{2}\left(1-\sin\frac{\theta}{2}\sin\frac{3\theta}{2}\right) \tag{3.131}$$

The K_{II} is the sliding mode stress intensity and can be obtained as following

$$K_{II} = \lim_{|\varsigma|\to 0} i\sqrt{2\pi\varsigma}Z_{II} \tag{3.132}$$

(c) Tearing Mode

$$K_{III} = \lim_{|\varsigma|\to 0}\sqrt{2\pi\varsigma}Z_{III} \tag{3.133}$$

The stress intensity factor is a fundamental quantity that governs the stress field near the crack tip. Several methods have been used for the determination of stress intensity factors as listed following:

(a) Theoretical method (Westergard semi-inverse method and method of complex potentials)

(b) Numerical method (Green's function, weight functions, boundary collocation, alternating method, integral transforms, continuous dislocations and finite element method)

(c) Experimental method (photoelasticity, holography, caustics)

Theoretical method is generally restricted to plates of infinite extent with simple geometrical configurations of cracks and boundary conditions. For more complicated situations one must result to numerical or experimental methods.

The stress intensity factor is one of the key parameters for characterizing stress field around crack, which can be used as the criterion for crack extension.

1.   Single mode criterion

The single mode criterion can be expressed as follows:

$$K_I \geq K_{IC}, \; K_{II} \geq K_{IIC}, \; K_{III} \geq K_{IIIC} \tag{3.134}$$

here K_{IC}, K_{IIC}, K_{IIIC} are the fracture toughness of I, II and III modes separately, which is also the inherent property of materials.

2.   Mixed mode criterion

The mixed mode criterion can be acquired using Ellipsoid Criterion:

$$\left(\frac{K_I}{K_{IC}}\right)^2 + \left(\frac{K_{II}}{K_{IIC}}\right)^2 + \left(\frac{K_{III}}{K_{IIIC}}\right)^2 \geq 1 \tag{3.135}$$

4.2. *J*-Integral Theory

The stress intensity factor can only be applied to small yield around crack tip, other appropriate parameters should be developed to evaluate the large fracture strength. Rice [11] introduced path independent line integral as the elastic-plastic parameter for characterizing the status of crack which also named as *J*-integral. Hutchinson [12] and Rice and Rosengren [13] showed that *J* uniquely characterizes crack tip stress and strains in nonlinear materials. Thus the *J* integral can be viewed as both an energy parameter and a stress intensity parameter. After that, many researchers investigate the *J*-integral which establish the theoretical foundation of the path independent *J*-integral and its use as a fracture criterion. Presently, the main efforts in the study of elastic-plastic fracture mechanics is building up the evaluating method on fracture strength using *J*-integral while the yield materials around crack tip can be considered as non-linear elastic materials.

As for crack in the nonlinear elastic continuum medium, Rice [11] found that the integral around crack tip is path independent and is given by:

$$J = \int_{\Gamma} \left(w dy - T_i \frac{\partial u_i}{\partial x} ds \right)$$ 　　　(3.136)

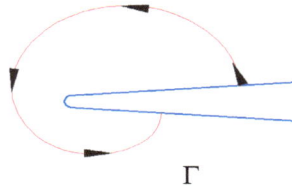

Figure 6: Arbitrary contour around the tip of a crack.

here *w* is the strain energy density, T_i is the component of the traction vector, u_i is the displacement vector component and *ds* is a length increment along the contour Γ. The stress energy density is defined as:

$$w = \int_{0}^{\varepsilon_{ij}} \sigma_{ij} d\varepsilon_{ij}$$ 　　　(3.137)

here σ_{ij} and ε_{ij} are the stress and strain tensors separately. The traction is a stress vector normal to the contour. That is, if we were to construct a free body diagram on the material inside of the contour, T_i would define the normal stress acting at the boundaries. The components of the traction vector are given by:

$$T_i = \sigma_{ij} n_j$$ 　　　(3.138)

here n_j is the component of the unit vector normal to Γ.

As for linear elastic materials, there some relationship as follows:

$$J = \frac{\kappa + 1}{8\mu} \left(K_I^2 + K_{II}^2 \right) + \frac{1}{2\mu} K_{III}^2 = G$$ 　　　(3.139)

As for nonlinear elastic materials, the system potential enclosed by curve Γ can be computed as follows:

$$\Pi = \int_{\Gamma} W(\varepsilon)dA_{\Gamma} - \int_{\Gamma} p_j u_j d\Gamma \qquad (3.140)$$

Therefore

$$J = -\frac{\partial \Pi}{\partial a} \qquad (3.141)$$

here a is the crack length. The J integral is essentially variation rate of system potential energy which is mainly transform into irreversible plastic work. If the work needed to extend crack a unit length is a constant, then the J integral based elastic-plastic fracture criterion can be deduced. It is because the J integral can be used to characterize the elastic plastic stress field solved by deformation theory that the J integral is selected as elastic plastic fracture criterion.

In 1968, Hutchinson [12], Rice and Rosengren [13] investigated the elastic plastic stress field around crack using deformation theory and acquired singular solution as follows:

$$\sigma_{ij} = \left(\frac{J}{a\varepsilon_Y \sigma_Y Ir} \right)^{\frac{1}{m+1}} \tilde{\sigma}_{ij} \qquad (3.142)$$

$$\varepsilon_{ij} = a\varepsilon_Y \left(\frac{J}{a\varepsilon_Y \sigma_Y Ir} \right)^{\frac{m}{m+1}} \tilde{\varepsilon}_{ij}(\theta) \qquad (3.143)$$

$$u_i = \left(\frac{J}{a\varepsilon_Y \sigma_Y Ir} \right)^{\frac{m}{m+1}} r^{\frac{m}{m+1}} \tilde{u}_i(\theta) \qquad (3.144)$$

here I is definite integral of θ, $\tilde{u}_i$ is a function of θ. In reality, it is difficult to solve the J integral using equation (3.142) ~ (3.144) because of the complex regular expression of $\tilde{\sigma}_{ij}$, $\tilde{\varepsilon}_{ij}$ and u_i. The numerical method and the energy method are the two practical solutions. The numerical method mainly makes use of elastic-plastic finite element method and integrates along several paths around crack tip and acquires the J integral. The final J integral can be computed as follows:

$$J = \sum \frac{J_i}{n} \qquad (3.145)$$

here J_i is the J integral corresponding to path Γ_i, n is the number of integrate path. The integrate path is generally continuous smooth curve which can reduce the error resulted by the discontinuous surface force.

4.2.1. Geometrical Criterion

The geometrical criterion mainly takes effect through judging if the geometrical size of materials exceeding the criterion. Fig. (**7**) shows the geometrical model in which a separation line is defined. The nodes at the chip side and the nodes at

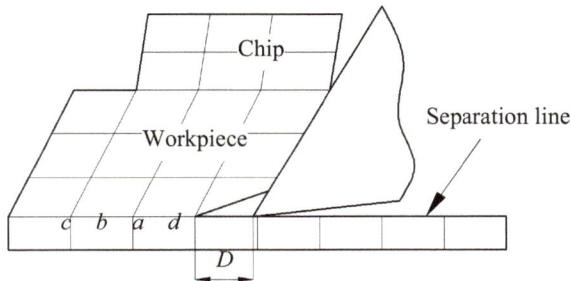

Figure 7: Geometrical criterion model.

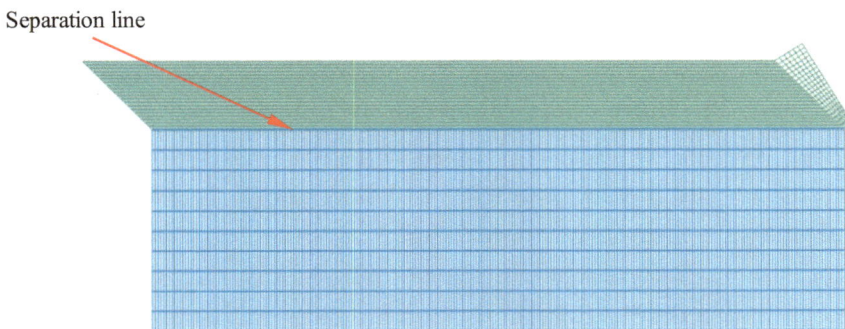

Figure 8: FE simulation based on geometrical separation criterion.

workpiece side are overlapped at the beginning. But the separation of two nodes occurs when the distance D between the tool cutting edge (point d, in Fig. (**7**)) and the node immediately ahead (node a) becomes less than a predefined critical value thus the machined surface and the chip bottom are generated.

Usui and Shirakashi [14] first put forward the geometrical criterion and found it is a stable criterion. Komvopoulos and Erpenbeck [15] pointed that there should be enough distance between tool tip and the overlap point to prevent the convergence problem resulted by the excessive distortion of finite element mesh. Zhang and Bagchi [16] brought forward that the geometrical distance should be less than 30 percent to 50 percent of element length. Furthermore, they also put up a new geometrical separation criterion which is based upon the ratio of geometrical distance to depth of cut which is equivalent to the microscopic fracture mechanics criterion.

The geometrical criterion is simple to be used in the FE computation. However, the distance (D) between tool tip and the separation point is closed to zero which result in the difference between the set value of D with the reality. The selection value of D will have a great influence upon the convergence of FE simulation and only the experienced researcher can deduce appropriate valuable critical value. In addition, the separation line which separates the mesh of chip and that of the workpiece should be built up in advance. Fig. (**8**) shows the FE simulation of precision machining process based on geometrical separation criterion.

5. FINITE ELEMENT MODEL ABOUT MNT

The FEM mesh is constituted by elements that cover exactly the whole of the region of the body under analysis. These elements are attached to the body and thus they follow its deformation. Metal cutting process is a large deformation and finite strain related elastic-plastic process. Therefore, both nonlinear material property and the nonlinear geometry property ought to be considered in the numerical analysis. Presently, typical finite element formulations used in metal cutting include Lagrangian or Eulerian method. Lagrangian formulation bases upon the original geometry which also termed as particle coordinates description, Eulerian formulation bases upon the deformed geometry which termed as floating coordinate description. These formulations are particularly convenient when unconstrained flow of material is involved, *i.e.*, when its boundaries are in frequent mutation. In this case, the FE mesh covers the real contour of the body with sufficient accuracy. On the other hand, the Eulerian formulation is more suitable for fluid-flow problems involving a control volume. In this method, the mesh is constituted of elements that are fixed in the space and cover the control volume. The variables under analysis are calculated at fixed spatial location as the material flows through the mesh. This formulation is more suitable for applications where the boundaries of the region of the body under analysis are known a priori, such as in metal forming.

Although both of these formulations have been used in modeling metal cutting processes, the Lagrangian formulation is more attractive due to the ever-mutating of the model used. The Eulerian formulation can only be used to simulate steady state cutting. As a result, when the Lagrangian formulation is used, the chip is formed with thickness and shape determined by the cutting conditions. However, when one uses the Eulerian formulation, an initial assumption about the shaped of the chip is needed. This initial chip shape is used for a matter of convenience, because it considerably facilitates the calculations in an incipient stage, where frequent problems of divergence of algorithm are found.

The Lagrangian formulation, however, also has shortcomings. First, as metal cutting involves severe plastic deformation of the layer being removed, the elements are extremely distorted so the mesh regeneration is needed. Second, the node separation is not well defined, particularly when chamfered and/or negative rake cutting edge tools are involved in the simulation. Although the severity of these problems can be reduced to a certain extent by a denser mesh and by frequent re-meshing, frequent mesh regeneration causes other problems.

These problems do not exist in the Eulerian formulation as the mesh is spatially fixed. This eliminates the problems associated to high distortion of the elements, and consequently no re-meshing is required. The mesh density is determined by the expected gradients of stress and strain. Therefore, the Eulerian formulation is more computationally efficient and suitable for modeling the zone around the tool cutting edge, particularly for ductile work materials. The major drawback of this formulation, however, is that the chip thickness should be assumed and kept constant during the analysis, as well as the tool-chip contact length and contact conditions at the tool-chip and tool-workpiece interfaces. As the chip thickness is the major outcome of the cutting process that defines all other parameters of this process so it cannot be assumed physically. Consequently, the Eulerian formulation does not correspond to the real deformation process developed during a real metal cutting process.

The Lagrangian formulation under finite deformation is as follows:

$$\left\{ p_t^\alpha \right\} = \iiint_{V_0} \left[B_{ij}' \right]^T S_{ij} dV_0 + \iiint_{V_0} \left[B_{ij}'' \right]^T S_{ij} dV \tag{3.146}$$

where $\left\{ p^\alpha \right\}$ denotes the column vector of external force exerted at the discrete element nodes, $\left[B' \right]$ is the geometry matrix in the case of finite strain conditions and the $\left[B'' \right]$ is the additional item induced by the geometric nonlinear conditions.

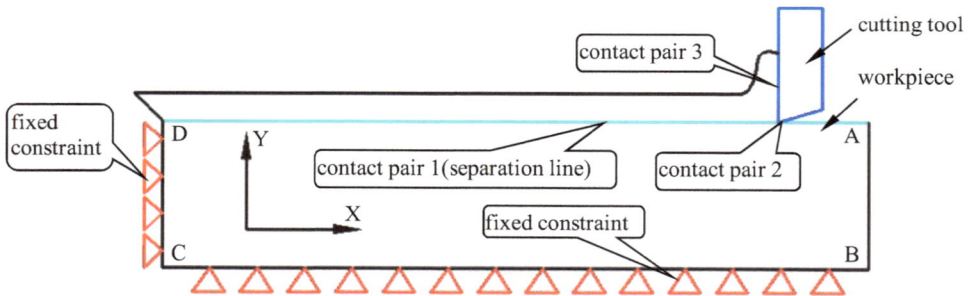

Figure 9: Schematic of FE model.

Figure 10: Finite element mesh of workpiece, chip layer and cutting tool.

In this research, the finite element analysis is utilized to gain good understanding of the materials deformation behavior underlying machining of titanium alloy. Among the different alloys of titanium, Ti-6Al-4V is by far the most popular with its widespread use in the chemical, surgical, ship building and aerospace industry. The primary reason for wide applications of this titanium alloy is due to its high strength-to-weight ratio that can be maintained at elevated temperatures and excellent corrosion and fracture resistance. On the other hand, Ti-6Al-4V is notorious for poor machinability due to its low thermal conductivity that causes high temperature on the tool face, strong chemical affinity with most tool materials, which leads to premature tool failure, and inhomogeneous deformation by catastrophic shear that makes the cutting force fluctuate and causes tool wear, thereby aggravating tool-wear and chatter. This poor machinability has limited cutting speed to less than 60 m/min in industrial practice. Numerical analysis of Ti-6Al-4V machining process using finite element method is of great importance on understanding the physical essence and optimizing the machining technique parameters (Table **1**).

Table 1: FEM simulation parameters

Material	Titanium
Size (*mm*)	$0.4 \times 0.15 \times 0.1$
Physical Property	Elastic-Plastic Solid
Depth of Cut(μm)	35
Speed(*mm/s*)	200
Temperature($^\circ C$)	20

A schematic diagram of the FE model is shown in Fig. (**10**). A rigid cutting tool moves towards the left with a constant velocity and removes a layer of workpiece material. The thickness of layer to be removed is decided a priori and it is assumed that the materials separate along the line of cut, namely, the path traced by tip of tool. To facilitate this material separation several contact pairs are defined in the FE model which is illustrated in Fig. (**9**).

❖   Contact pair 1 is defined between the bottom surface of the chip layer and the surface of the workpiece along the line of separation.

❖   Contact pair 2 is between the workpiece and tool tip.

❖   Contact pair 3 is between the chip and rake face of tool.

The materials properties are the same across the contact surface in the contact pair 1, namely, between the chip layer and the rest of the workpiece. These two surfaces remain in contact until a certain condition-a chip separation criterion is satisfied. Two common chip separation criterion are illuminated in the above section. The tool is modeled as rigid with an artificially high modulus of elasticity. Plane strain element is used in this numerical analysis and all nodes in the workpiece and tool have two displacement degrees and one temperature degree of freedom. The initial orientation of chip layer elements is used to alleviate the numerical problems due to distortion of the elements as they separate from the workpiece and slide along the rake face of the tool. An initial chip separation is adopted in order to achieve a smooth transition of the cutting process in the beginning when the tool first comes in contact with the workpiece. The extra triangle of the chip layer at the left end is used simply to make the mesh generation simpler and will not affect the simulation results. Since the tool is modeled as rigid, its size is not a critical issue while it ought to be sure that the length of the rake face is large enough to accommodate the flowing chip. The thermal properties of the tool are of important since it is thermo-mechanical coupled analysis. In particular, specific heat and conductivity of tool material affect the heat flow in the problem and thus influencing the nature and magnitude of temperature field.

The boundary conditions for the chip-workpiece-tool system are as follows:

1.   Nodes along the edges AB, BC, CD are fixed in the X and Y directions.

2.   The upper boundary of the tool moves incrementally towards the left with uniform velocity while it is restrained vertically.

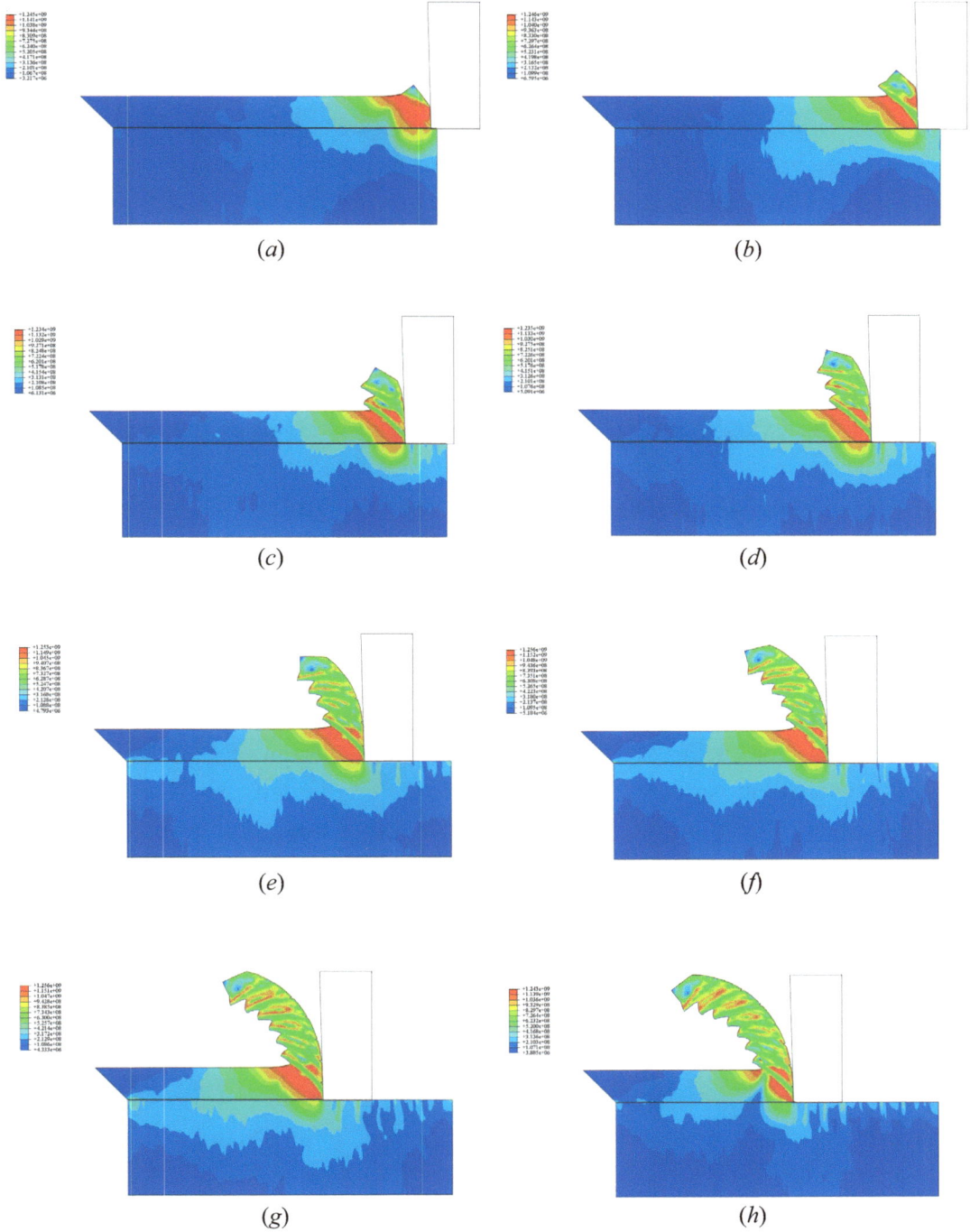

Figure 11: Finite element simulation results.

6. TRIBOLOGY BEHAVIOR AND CHIP FORMATION AT TOOL-CHIP INTERFACE (FIG. 11)

In this investigation, the Eulerian Formulation is used to mimic tribology behavior of the tool-chip and the tool-workpiece interfaces in the precision machining process. The tool-chip and the tool-workpiece interfaces are those contact areas that directly participate in the cutting process. Considerably different functions of the tool-chip and tool-workpiece interfaces in this process define the differences in the tribological processes at these interfaces. Although there are a number of differences, there are also a number of similarities in the tribology of these interfaces as high contact pressures and temperatures, contact with freshly formed counter-surfaces and high sliding velocities. There is also a correlation between the stresses at these interfaces as they both are related to the same state of stress in the deformation zone. As such, however, the tool-chip interface plays a leading role that defines to a large extent this state. Therefore, a greater part of the considerations presented in this section is dedicated to the tribological conditions at the tool-chip interface.

The main objective of this section is to discuss the tribological conditions at the tool-chip and tool-workpiece interfaces. The understanding of these conditions and the proper utilization of this understanding in the design of the cutting process should result in the increased efficiency of the cutting system and in the reduction of tool wear. These results can be used in the meaningful selection of the machining regime, tool geometry and tool materials (including coatings).

6.1. Tool-Chip Interface

When a material is cut, the cutting force acts mainly through a small area of the rake face, which is in contact with the chip and thus, is known as the tool-chip interface. Therefore, it is of interest in determining the cutting force, developing the theory of tool wear and understanding the mechanics of chip formation to establish the tribological characteristics of the tool-chip interface.

The main tribological characteristics of the tool-chip interface are:

❖   The contact length: the length of the tool-chip contact

❖ The sliding velocity

❖ The friction force at the tool-chip interface

❖ The specific frictional force which is the mean shear stress

❖ The normal force at the tool-chip interface

❖ Mean normal stress at the tool-chip interface

❖ Mean contact temperature at the tool-chip interface

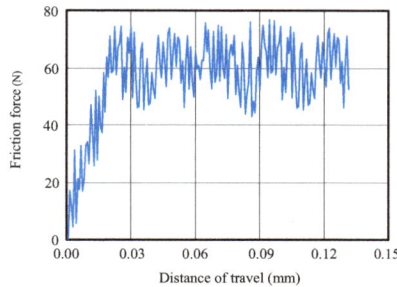

Figure 12: Variation of friction force obtained from FE simulation.

6.2. Friction Coefficient

In Merchant's analysis, it is implied that the contact between the tool and the chip is a sliding contact, where the coefficient of friction is constant. In most engineering and physical situations, friction effects at a tribological interface are described by a constant coefficient of Coulomb friction μ_f,

$$\mu_f = \frac{F}{N} \tag{3.147}$$

here N is the normal force acted upon the considered interface, F is the friction force at this interface.

Although it is well established that contact between the two bodies is limited to only a few microscopic high points (asperities), it is customary to calculate the stresses by assuming that the forces are distributed over the total (apparent) contact area. Such an approximation, however, is not far from reality in machining, where the actual and apparent contact areas are practically the same due to high contact pressures. If it is so, the numerator and denumerator of Equation (3.147) can be divided over the tool-chip contact Ac and then rescaling that the mean normal stress at the interface is $\sigma c = N/Ac$ and the mean shear (frictional) stress at the interface is $\tau c = F/Ac$, one can obtain

$$\mu_f = \frac{\tau_c}{\sigma_c} \tag{3.148}$$

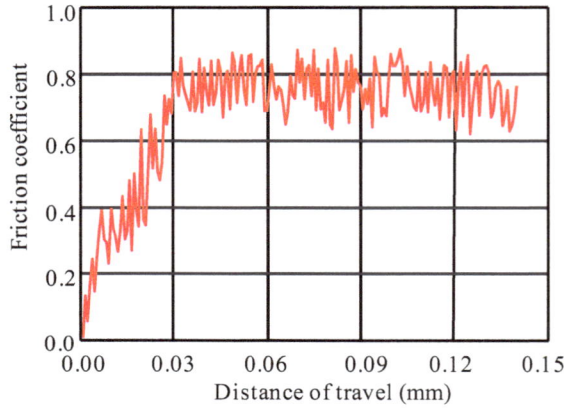

Figure 13: Variation of friction coefficient obtained from FE simulation.

The above analysis was for sliding friction at the interface, as in our first encounter with friction in elementary physics. At the other extreme, we can envision a situation where the interface has constant film shear strength. The most usual case is sticking friction, where there is no relative motion between the chip and the tool at the interface. For sticking friction, this shear strength should be equal to the flow stress in shear, k_c, and the normal stress to the yield stress of the work material σ_y. Using the *Von Mises* yield criterion, the coefficient of friction under sticking conditions is

$$\mu_f = \frac{k_c}{\sigma_y} = \frac{\sigma_y / \sqrt{3}}{\sigma_y} = 0.557 \tag{3.149}$$

Therefore, the value of the friction coefficient μ_f defined by Equation (3.149) should be considered as the limiting value so that if $\mu_f \geq 0.577$, no relative motion can occur at the interface.

In the practice of materials machining, this is not the case. Many reported values of μ_f obtained in metal cutting tests are well above 0.577. Fig. (**13**) gives the FE simulation result of friction coefficient at the tool-chip interface which is also above 0.577 as the experimental study. The friction coefficient increases quickly at the initial stage of cutting which is also called unsteady cutting. After that, the friction coefficient gradually approaches to a stable value which is corresponding to the steady cutting. It is true that, in general, the coefficient of friction for sliding surfaces remains constant within wide ranges of the relative velocity, apparent contact area and normal load. In contrast, in metal cutting, the coefficient of friction varies with respect to the normal load, the relative velocity and the apparent contact area.

The coefficient of friction in metal cutting was found to be so variable that it is doubted whether this term served any useful purpose. Furthermore, it is clear that the concept of the coefficient of friction is inadequate to characterize the sliding between chip and tool and thus recommended to discontinue using the concept of the coefficient of friction in metal cutting. An extensive analysis of the inadequacy of the concept of the friction coefficient in metal cutting was presented by Kronenberg [17] who stated "I do not agree with the commonly accepted concept of coefficient of friction in metal cutting and I am using the term 'apparent coefficient of friction' wherever feasible until this problem has been resolved." Unfortunately, it has never been resolved although more than 40 years have passed since this statement of Kronenberg.

To model friction conditions in metal cutting and thus to determine the real value of the friction coefficient, a large number of theoretical analysis have been carried out to determine the influence of various parameters on the interfacial friction. Despite the improvements made in the modeling of friction in machining processes over the last 30 years, there are still a number of limitations so that the results of these studies are hardly applicable to metal cutting. There are some severe drawbacks of the discussed theoretical considerations are:

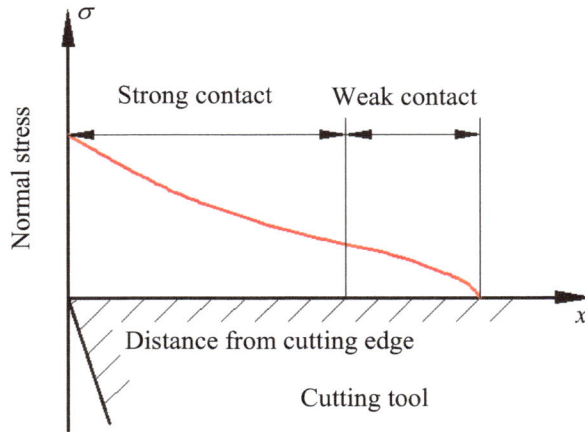

Figure 14: Stress distribution at the tool-chip interface.

- At the interface, the soft material is always assumed to be rigid, perfectly plastic, incompressible and isotropic at all stages of deformation. Moreover, no superficial and in-depth residual stresses due to previous deformation can be accounted for. Obviously, this is not the case in metal cutting and particularly at the tool-chip interface.

- The approaches known so far consider only two-dimensional asperity interaction ignoring the three-dimensional nature of asperities and actual multi-asperity interaction. The shape of the asperity should be well defined so that its initial and final point should lie on the same line parallel to the direction of relative motion. Moreover, this shape determines the slip-line field where the discontinuity of tangential velocity leads to infinite calculated strains. No wonder the chip formation model predicts an increase in the coefficient of friction with improved lubrication.

- The sliding speed was not considered to be an important factor. The temperature and strain rate effects are poorly accounted for. The sliding speed even in the verification experiments was 0.3 mm/s and this is obviously way below the sliding speeds found at the tool-chip interface. It was also found that the strains determined from the experimental flow fields are significantly lower than those calculated from the model.

6.3. Contact Stress Distribution

This section aims at obtaining a better understanding of the conditions at the tool-chip interface by studying the distribution of the normal and shear stress at this interface (Fig. (**14**)). The contact stress decrease quickly away from the cutting edge. The entire tool-chip contact area is divided into two parts, one area with larger normal stress named strong contact area while the other area with small normal stress named weak contact. The demarcation point is set according to inflection point of the shear stress of this area. The existence of weak contact area ought to be result of curling of chip. A variety of experimental techniques including photo-elastic tools, split tool dynamometer, transparent tool for the direct observation of the tool-chip interface, metallurgical examination of "quick-stop" chip-section including experimental slip line field method have been developed. The normal stress is aero at the point of chip separation and increases exponentially towards the cutting edge.

7. RESIDUAL STRESS (STRAIN) AT THE MACHINED SURFACE

The stress status of the machined surface such as residual stress and residual strain has a great influence upon quality of the final part. The residual tension stress in the machined surface will degrade the fatigue strength of workpiece. It is necessary to predict the surface residual stress and the geometry accuracy in the reality. In the materials processing technology, the influence layer generated by the former working procedure always removed by the current procedure which induces residual stress and strain in the newly generated is called work hardening. The work hardening can cause the change of the cutting force and the shear angle in the dual machining process.

In the elastic-plastic finite element simulation, the successive unloading is used to acquire the residual stress and strain. The successive unloading is used by exerting a load $-\Delta\{P\}$ until all of the loads are released. The computation method and the procedure is same with loading process but the element in the unloading stage experienced plastic deformation to elastic deformation which induce different element transformation order. Some elements may under the state of loading in the unloading stage which makes it necessary to determine the stress state of the element. Generally speaking, the relationship between stress and strain ought to conform to materials property curve while the stress and strain ought to emerge as linear elastic spring-back relationship in the unloading stage. The solution process corresponding to these two situations is thus different accordingly. The strain-hardening rate and the equivalent stress are very small in the case of large deformation stage and sometimes less than the computation error. In reality, the least square method is the effective method is used to accurately determine the stress status of element.

The material will be in the loading state if the material is isotropic and satisfy the Von-Mises yield criterion when the equivalent stress induced by the increasing of small displacement. It is simple to justify the loading status of the element if the increment of the equivalent stress is larger but the enough precision cannot be acquired if the increment of equivalent stress is small. The justification of unloading status by simple integration cannot assure enough precision and the computation of slope of effective stress-strain curve using normal equation is preferred:

$$A\sum X_i^2 + B\sum X_i = \sum X_i Y_i \qquad\qquad (3.150)$$

S, Mises

```
+9.689e+08
+8.906e+08
+8.123e+08
+7.340e+08
+6.557e+08
+5.774e+08
+4.991e+08
+4.208e+08
+3.425e+08
+2.642e+08
+1.859e+08
+1.077e+08
+2.935e+07
```

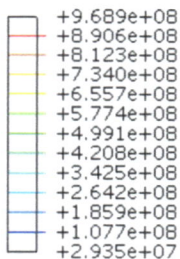

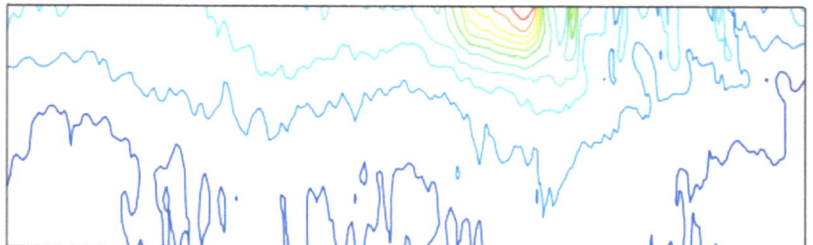

Figure 15: Residual stress in the machined surface.

$$A\sum X_i + BN = \sum Y_i \tag{3.151}$$

The slope acquired by the above equations can thus be used to determine the stress status of element. The residual stress distribution in the machined surface using above technique is shown in Fig. (**15**). A zone of plastic deformation extends underneath the machined surface. The depth of subsurface plastic deformation is found to be nearly equal to the depth of cut. This subsurface deformation results in intensive stress in the machined surface. Though the stress patterns shown are those with the load applied by the tool still present, elastic recovery caused by the unloading of the tool is not expected to significantly change the stress distribution close to the free surface. So the stress in the machined surface sufficiently far away from the tool can be taken to be the residual stress.

8. MICROSCOPIC ANALYZING THE MECHANISM OF MNT

It has been known for a very long time that the size effect exists in metal cutting, where the specific energy increases with decrease in deformation size. In the case of micro & nano-machining technique, most of the materials removal (plastic deformation) is accomplished by the tool tip as shown in Fig. (**16**). There is a substantial increase in the specific energy required with the decreasing of chip size.

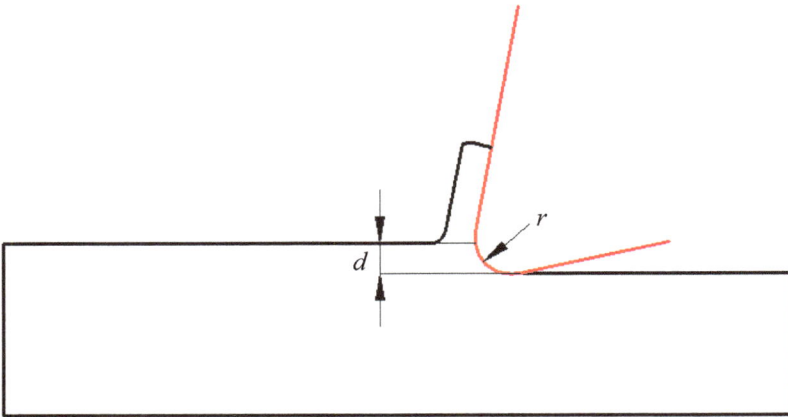

Figure 16: Illustration of ultraprecision cutting.

Also the small depth of cut increases the materials deformation and friction at the tool-chip interface and weakening the effect of materials removal which may influence the chip formation. Presently, the single-shear plane model is still used for studies on metal cutting. This traditional single-shear plane model on the chip formation is not applicable in MNT which is shown in Fig. (**17**).

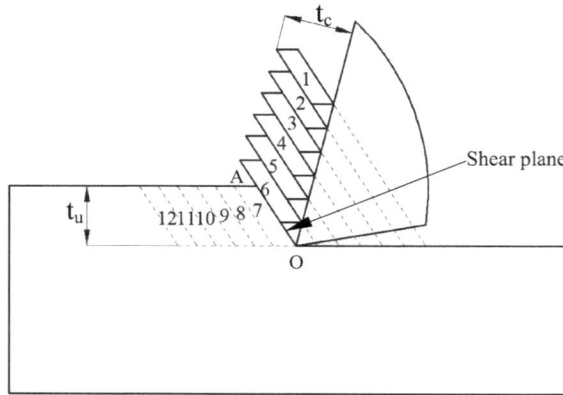

Figure 17: Single shear plane model of cutting

The single-shear plane model of chip formation has been constructed using simple observations of the metal cutting process at the end of the nineteenth century. The observations seem to have led to an idealized picture, which is known today as the single-shear plane model for orthogonal cutting, shown schematically in Fig. (**17**). The tool removes a stock of thickness t_u by shearing it ahead of the tool in a zone which is rather thin compared to its length, and thus it can be represented reasonably well by the shear plane OA. The position of the shear plane is defined by the shear angle φ. After being sheared, the layer being removed becomes the chip having thickness t_c, which slides along the tool rake face. The first orthogonal model was brought forward in 1937 by Piispanen [18] and termed as card model. In this model, the material cut is assumed as a deck of cards inclined to the cutting direction. Merchant assumed the chip to be formed over an infinite thin plane called shear plane. This shear plane starts from the cutting edge of the tool and crosses the chip on an angle with the cutting direction, which is termed as shear angle. When the chip passes the shear plane it is sheared away from the workpiece and increases in thickness. In this simulation, no single shear plane is observed in the whole precision machining process. On the other case, there some maximum stress band is continuously generated in front of cutting tool. This shear band possesses irregular geometry shape which extends from first deformation region to third deformation region.

A zone of plastic deformation extends underneath the machined surface. This subsurface deformation will result in compressive stresses in the machined surface. Though the stress patterns are those with the load applied by the tool still present, elastic recovery caused by the unloading of the tool is not expected to significantly change the stress distribution close to the free surface. The location of the nodes along the machined surface when compared with the location of tool cutting edge yields information about the elastic recovery of the machined surface after it passes under the tool. The elastic spring-back of the machined surface is found to be far less than the radius curvature of cutting edge which justify that most of the material in front of the rounded cutting edge is actually pushed ahead of the tool and not into the machined surface.

The simulation results also shows that the continuous internal curling chip is generated under current working conditions. At the beginning, part of chip adjacent to the tool tip begins to curl and form helix circle with small radius. After that, the larger helix circle surround the previous small one is gradually formed with the feeding of the cutting tool. The deformation coefficients ($\xi = \dfrac{t_c}{t_u}$) is gradually increased in this process which result in the increasing of cutting force. The stress along the free surface (back) of chip is tensile. It is also tensile along the surface of chip which has moved out of the contact with the tool rake face (front) while the σ_{yy} in the middle of the chip is compressive. Such a distribution of stress is the critical factor to develop initial formation of chip. Presently, the hypotheses propounded by various researchers to explain the curvature of the chip include (i) The cutting moment causes the chip to bend; (ii) The 'crushing' of chip in the secondary shear zone and the resultant acceleration of the work material in moving through the secondary shear zone causes the chip to lengthen along this side (the front side). This can also results in a curvature of the chip which is similar to the curvature of a bimetallic strip; (iii) The shear plane is curved in such a way that the shear plane angle is smaller near the exit of shear plane. Thus the chip velocity on the back side is smaller than the average chip velocity which causes the chip to curl.

The bending moment on the chip considered as a beam would result in compressive stress along the free surface (back) of the chip if hypothesis (i) was true. Crushing of the chip in the secondary shear zone will result in compressive σ_{yy} in the front (underside) of the chip. Only a curved shear plane would result in a stress distribution similar to that given by the finite element analysis, while simultaneously accounting for curl of the chip. It should be noted that though the chip does accelerate (due to secondary shear) as it flows along the rake face of tool, this is just an accessory to chip curl and not the cause of chip curl. The reason for the curvature of the shear plane can be found from a detailed analysis of the stress distribution in the zone of plastic deformation. Work in this direction is in progress.

REFERENCES

[1]   X.S. Han, "A study on the mechanism of ultraprecision manufacturing and its simulation", M.S. thesis, Tianjin University, Tianjin, China, 2002.

[2]   L.J. Segerlind, *Applied finite element analysis*. New York: John Whiley and Sons, 1984.

[3]   Von Mises, "Mechanics on plastic deformation in solids", Math. Phys., vol. 1, pp. 582-592, January 1913.

[4]   R. Hill, "New horizons in the mechanics of solids", J. Mech. Phys. Solids., vol. 5, pp. 66-71, May 1956.

[5]   F. W. Taylor, "On the art of cutting metals", Annual Meeting of the American Society of Mechanical Engineers, 1982, pp. 14-19.

[6]   J.T. Nicolson, "Machine Tool Design", *The Engineer*, vol. 99(385), pp. 331-332, April 1905.

[7]   J. S. Strenkowski, J. T. Carroll, "A finite element model of orthogonal metal cutting", *ASME J. Engi. for Indus.*, vol. 107: pp. 349-354 July 1985.

[8]   T.L. Anderson, *Fracture mechanics - fundamentals and applications*. New York: CRC Press, 1995.

[9]   E.E. Gdoutos, *An introduction to fracture mechanics*. Dordrecht: Kluwer Academic Publishers, 1993.

[10]   G. R. Irwin, *Fracture, Encyclopedia of Physics, VI (Elasticity and plasticity)*. New York: Springer, 1972.

[11]   J. R. Rice, "A path independent integral and the approximate analysis of strain concentration by notches and cracks", J. Appl. Phys., vol. 35, pp. 379-386, February 1968.

[12]   J.W. Hutchinson, "Singular behavior at the end of a tensile crack tip in a hardening material", J. Mech. Phys.Solids, vol. 16, pp. 13-31, August 1968.

[13]   J. R. Rice, G. F. Rosengren, "Plane strain deformation near a crack tip in a power law hardening material", J. Mech. Phys. Solids, vol. 16, pp. 1-12, March 1968.

[14]   E. Usui, T. Shirakashi, "Mechanics of machining - From descriptive to predictive theory, On the art of cutting metals - 75 years later a tribute to F W Taylor", ASME PED-7, 1982, pp. 13-30.

[15]   F. K. Komvopoulos, S. A. Erpenbeck, "Modeling of orthogonal metal cutting", Trans of ASME J. Eng. Ind, vol. 113, pp. 253-267, April 1991.

[16]   B. Zhang, A. Bagchi, "Finite element simulation of chip formation and comparison with machining experiment, Computational method in material processing", ASME publication - PED, vol. 61, pp. 61-74, January 1992.

[17]   M. Kronenberg, *Machining Science and Application-Theory and Practice for Operation and Development of Machining Processes*. Oxford: Pergamon Press, 1966.

[18]   M.C. Shaw, *Metal Cutting Principles*. Oxford: Oxford Science, 1984.

Send Orders for Reprints to reprints@benthamscience.net

CHAPTER 4

Multiscale Analysis of MNT

Abstract: MNT is a technology that accurately produces geometrically dimensional shapes in the micro & nanometer scale. An emphasis on micro & nanoscale entities will make our manufacturing technologies and infrastructure more sustainable in terms of reduced energy usage and environmental pollution. The minimizing of the workpiece and the small depth of cut makes the materials removal or deformation in the MNT different from the traditional machining technique. Materials properties follow from their atomic and microscopic structures and exhibit different properties at different scales. Bulk materials of micro & nanometer scale are relatively large and may be expected to behave as "macroscopic" object in some respects. At the same time they are small enough to permit the long-time simulations necessary for investigation of microscopic properties such as self-structural organization processes. It is difficult to deeply investigate materials deformation or failure mechanism using single macroscopic or phenomenological scale method. Multiscale method can study material behavior at different length scale and temporal scale simultaneously which can uncover important properties and materials response in MNT from atomic to microscopic to mesoscopic to macroscopic scales. Multiscale method offer the best hope for bridging the traditional gap that exists between experimental approach, the theoretical approach and computational modeling for studying and understanding the deformation and removal mechanism of materials in MNT. Multiscale method conform to the basic natural philosophy ideas, namely, every things should evolve from quantitative change to qualitative change. Owing to the central role that multiple scale methodology appears poised to play in the computational mechanics and materials science in the foreseeable future, this chapter introduces multiscale method and some recent applications.

Keywords: Multiscale method, coupling continuum and atomistic method, transition model, geometrical-physical model, multiscale simulation of microcutting.

INTRODUCTION

MNT are processes which a very small amount of materials are removed (per cutting edge, for example) and for which the surface or feature created can be characterized by stringent tolerances on form, dimension or surface characteristics. The MNT processes include machining with very small chip thickness also the process in which no discernible "chip" is produced but features

Xuesong Han

are created. These processes include single point diamond turning in which most of the cutting work accomplished by the tool tip, abrasive machining (including lapping, polishing and honing) and can be characterized by either two body or three body abrasive interactions. Furthermore, there are many nontraditional processes include atomic bit processing with electron beams or evaporation such as AFM and STM processing. There are various ways to classify precision material removal processes. We have presented one above, based on the "uncut chip thickness", and will discuss this in more detail. As we shall see, there can often be two or more processes active in a particular precision material removal operation. For example, Chemical Mechanical Polishing is, in fact, a combination of mechanical abrasive action and chemical corrosion. Our interest in precision material removal derives from its importance in the manufacture of products and devices with the most stringent tolerances of products and associated computer peripherals. There are some common features among all of the specific techniques of MNT, namely, the material removal volume or the depth of cut or the radius of cutting tool is very small, usually fall into nanometer or micrometer length scale. Experimental investigation is time consuming and expensive, model based numerical simulation provides an effective alternative for solving such problems. By using such techniques detailed information on the state of material undergoing deformation is determined and more realistic conditions during machining can be incorporated.

Traditional continuum mechanics use experimental results as their guide to choose the appropriate constitutive behavior, but recent trends toward modeling phenomenon happened at small scale mean that many material properties of interest cannot be obtained from direct experimental observation. Atomistic models provide a means to accurately describe materials behavior on the scale of angstrom, but are severely limited in that they can only model a small number of atoms. Therefore, the mechanism of MNT cannot be modeled by pure continuum methods, because they are too small, nor can they be entirely modeled by molecular methods because they are too large. Coupled multiscale methods are urgently needed for this class of problems. Owing to the combination of constantly increasing computational power and the increased knowledge and understanding of material behavior, multiscale modeling methods have recently emerged as the tool of choice to link the mechanical behavior of materials from the smallest scale of atoms to the largest scale of structures. Multiscale methods offer the best hope for bridging the traditional gap that exists between experimental approach, the theoretical approach and computational modeling for studying and understanding the behavior of materials. When we say that a problem in science is "multiscale" we broadly mean that it involves phenomena at disparate length and/or time scales spanning several orders of magnitude. More importantly, these phenomena all play key roles in the problem, so that we cannot correctly model the interesting behavior without explicitly accounting for these different scales. Indeed, the message we have tried to carry throughout this chapter is that there is a need to model materials at many scales, and to make connections between them. However, when we speak of multiscale modeling, we tend to be referring to something more specific, meaning that the problem is tackled with a conscious effort to span multiple scales simultaneously. In many cases, multiscale methods involve just two scales, a "coarse scale" and "fine scale", each of which plays a role in the problem. Depending on the perspective, multiscale models offer different advantages. For the fine-scale modeler, a multiscale approach allows one to study much larger systems (or longer times) than could be studied using the fine scale alone. On the other hand, the coarse-scale expert views the multiscale model as a way to establish the constitutive laws of the problem from first principles, or at least from a more fundamental scientific basis than could be realized from the coarse-scale alone. In the former case, the advantage is clearly a practical one. In the latter case, the advantage is often better scientific insight into the why of a particular phenomenon, but there is also a practical benefit. For example a constitutive law may be provided in cases where direct experimental determination of the coarse-scale behavior is not possible. Understanding of the fundamental physical mechanisms behind MNT is a major scientific task and it is the reason why multiscale analysis is so important. This is based on the judgment that multiscale investigates the effects of a hierarchy of internal structures on material behavior, thus making it realistic to understand fundamental mechanisms and to avoid misunderstandings based on macroscopic observations and resulting assumptions. This becomes possible due to the fact that the rapid development of computer and computational algorithms makes modeling and simulation an increasingly powerful tool. Traditionally, physics takes theory and experiment as its two fundamental bricks. In many cases, theory is developed based on simplified assumptions and can only be verified with limited tests. Multiscale simulation sits in between these two, which makes the development of theory more realistic and even allows "thought test" or tests which cannot be done by experiment. Furthermore, multiscale simulation makes the content of experiment rich and deep, reducing its uncertainty and making it easier to be intrinsically connected to or be explained by the theory. Because it starts from the fundamental scale and can be calculated quantitatively, multiscale analysis can bring new concepts, methods, mechanisms and phenomena for, say, upgrading or designing materials. These may not be possible or at least need a longer time to develop when one stays only at the phenomenological stage. The significance of multiscale analysis in science and technology is thus defined and illustrated.

Multiscale methods generally imply the utilization of information at one length scale to subsequently model the response of the material at larger length scales. These methods can be divided into two categories: hierarchical and concurrent. Hierarchical multiple scale methods directly utilize the information at a small length scale as an input into a larger model *via* some type of averaging process. In this method, the response of representative volume element at the fine scale is first computed over a range of expected inputs and from this the stress-strain law is extracted. The stress-strain law can be a constitutive equation with parameters determined by the fine scale solutions. For the linear response, this process can be simplified since there is a robust theory homogenization whereby the linear constants can be effectively extracted. The Young's modulus is a good example of this; the structural material stiffness is found as a single quantity, through homogenization of all defects and microstructure at the micro- and nanoscales. For severely nonlinear problems, hierarchical models become more problematical, particularly if the fine scale response is path dependent. Hierarchical models are invalid and cannot be used when the failure occurs in many circumstances.

Concurrent multiscale methods are those that run simultaneously; in these methods, the information at the smaller length scale is calculated and inputted into the larger scale model on the fly. Both compatibility and momentum balance are enforced across the interface. For example, the MAAD (macroscopic, atomistic, ab initio dynamics) method, the bridging domain method use quantum mechanics models (in the bond-breaking subdomain) which linked to atomistic and continuum models belong to concurrent multiscale methods. Such models are quite effective when the subdomain, where a higher order theory is required, is small compared to the domain of the problem. In the study of fracture, fine scale models can be inserted in hot spots where the stress or strain becomes large. There is a class of semi-concurrent method in which a fine-scale model is calculated concurrently with coarse-grained model. They distinguished from concurrent methods by the fact that the fine-scale model is not coupled directly with the coarse-grained model where compatibility and momentum balance are only satisfied approximately. The reduction in computational effort from a fully concurrent approach is not great. The appeal of this method lies in the fact that the fine-scale and the coarse-scale models can be run by separate software and the avoidance of numerical problems associated with rapid changes in element sizes.

Despite significant developments in materials simulation techniques, the goal of predicting reliably the properties and behaviors of in materials processing has not yet been achieved. This situation exists for several reasons that include a lack of full understanding of material behavior at different scales, absence of scaling laws, computational limitations, and difficulties associated with experimental measurements of material properties at micro and nano scales. For example, the information on the mechanical behavior of materials at nano level is not presently available as input to nanotechnology for the manufacturing of nanocomponents or microelectromechanical systems (MEMS). Scaling laws governing the mechanical behavior of materials from atomistic (nano), *via* mesoplastic (micro), to continuum (macro) scales are very important to numerous applications. Scaling laws are also important for applications where two length scales of different orders of magnitude are involved. For example, one is atomistic (nano) and the other mesoplastic (micro) as in nanocutting, or, one is mesoplastic (micro) and the other continuum (macro) as in conventional indentation. Appropriate scaling laws may extend the extensive knowledge accumulated over time on material behavior at the macro (or continuum) level to the atomistic (or nano) level, *via* mesoplastic (micro) level. Micro and nanomechanics property of materials and the corresponding deformation behavior are of critical importance in understanding materials behavior in MNT as it allow simulation across different length and temporal scales.

1. QUASI-CONTINUUM (QC) METHOD

A well-known quasi-static multiscale method, the quasi-continuum method, was developed by Tadmor *et al.* (1996) [1]. The quasi-continuum (QC) method is a multiscale method combining atomistic simulations with nonlinear continuum finite elements. Examples of applications and further improvements on the quasi-continuum method are the works of Miller *et al.* (1998) [2] and Knap and Ortiz (2001) [3]. A recent review concentrating on the history and development of the quasi-continuum method is given by Miller and Tadmor (2003) [4]. While the quasi-continuum method is essentially an adaptive FE method, the atomistic to continuum link is achieved here by the use of the Cauchy-Born rule. It is a landmark work in atomistic/continuum coupling. It was one of the first works to introduce the concept of using the Cauchy-Born rule in the continuum part of the model, so that the two models are compatible. From a coupling point of view, the quasi-continuum method is basically a direct atomistic-continuum coupling scheme where the selected atoms are coincident with finite element codes. Therefore, as in any direct coupling method, the mesh size must be decreased to the atomistic lattice spacing wherever atomistic potentials are to be used, which complicates the mesh generation process.

1.1. Representative Atom and the Corresponding Crystallite

In the QC method, it is assumed that there is an underlying atomistic model of the material which is a good description of the material's behavior. The undeformed body is assumed to be a crystalline solid, *i.e.*, a collection of N atoms, which occupy a region and may be arranged in many grains. The undeformed position X_i of any atom i is obtained from the coordinate of a reference atom and an associated set of Bravais lattice vectors. Based on this concept, in the QC literature, only "representative atoms" or simply "repatoms" are taken to be part of the set of degrees of freedom for the solution. The set of representative atoms are selected as the set of atoms nearest to the element quadrature points located at the element centroids. There are two kinds of repatoms in QC: non-local and local. This terminology is used to distinguish between repatoms in the atomistic region and repatoms in the continuum region. In other words, the term "nonlocal repatom region" in the QC method should be regarded as the real atomistic region, while the term "local repatom region" is considered as the continuum or finite element region. The repatom in the local region in general is the node of FEM. Furthermore, the locations of any atom in the local region are determined by location of repatoms because they are kinematically constrained by these repatoms through interpolation functions. However, single atoms are chosen to represent two elements in the coarser region while many of their neighbors remain unrepresented. This leads to an increase in noncentrosymmetry effects and may result in small non-physical fluctuations in deformation in these transition areas. To minimize this effect, QC developers maximize the number of atoms in the solid by using some specific algorithm. The goal of QC statics methods is to find the atomistic displacements that minimize the total potential energy through approximating the energy by introducing these representative atoms to substantially reduce the degrees of freedom.

With the set of representative atoms defined QC construct the representative crystallites that corresponding atoms are embedded in. As the energy of atom depends on all neighboring atoms that lie within its cutoff region following deformation, it is important to recall that the representative crystallites are constructed in the undeformed region. The intensive task about classical energy computation of a given atom adopted in molecular dynamics is alleviated by constructing from a mathematical description of the perfect lattice in QC program and locating atom's neighbors is straightforward. The deformed positions of the atom are not stored but can be readily obtained by interpolating the FEM displacement field to their equilibrium sites. The representative crystallite is defined after the selection of representative atom which should ensure element size is sufficiently large to account for all atoms that move inside the cutoff sphere of representative atom. The deformed neighborhood is constructed in the

QC method by applying the continuum displacement field to its undeformed neighborhood. The necessary crystallite size will vary throughout the solid, growing larger where the deformation is severe and shrinking down to the cutoff sphere where the solid is undistorted. The representative crystallite radius computation in QC formulation is endowed with a built-in fracture mechanism. As the deformation progresses the representative crystallites of the element near the slip planes will grow. However, they cannot grow beyond the model boundaries and hence eventually elements near the surface will start breaking which is in identical with real physical process. The representation is of varying density with more atoms sampled in highly deformed regions (such as near defect cores) and correspondingly fewer in the less deformed regions further away, and is adaptively updated as the deformation evolves. The concepts of representative atom and representative crystallite are the fundamental issues of quasi-continuum methodology which relates with energy computation inside continuum model.

1.2. Formulation of Total Potential Energy in QC

Based on the kinetic model environment, the energy of individual representative atom can be computed either in nonlocal fashion in correspondence with straightforward atomistic methodology or within a local approximation as befitting a continuum model. Once the atomic positions have been given, from the standpoint of a strictly atomistic perspective, the total energy function E^{tot} determined from the relative positions of all the atoms in the system (whose total number is n) can be expressed:

$$E^{total} = \sum_{i=1}^{n} E_i \left(\vec{x}_1, \vec{x}_2, \cdots\cdots, \vec{x}_n \right) \qquad (4.1)$$

Where $\vec{x}_i$ is the position of atom i.

The deformed configuration of the body is described by a displacement function $\vec{u}$ which depends on X and the deformed position of any atom i can be obtained as

$$\vec{x}_i = \vec{X}_i + \vec{u}_i$$
$$\vec{u}_i = \vec{u}(\vec{X}) \qquad (4.2)$$

On loading the solid, in addition to potential energy there is work potential due to external loads applied to atoms. Thus, the total potential energy of the system (atoms plus external loads) can be written as

$$\Pi(\vec{u}) = E^{tot}(\vec{u}_1, \vec{u}_2, \cdots\cdots, \vec{u}_n) - \sum_{i=1}^{n} \vec{f}_i \cdot \vec{u}_i \qquad (4.3)$$

where E^{tot} is the total energy of the system obtained from an atomistic formulation, $\vec{f}_i$ is the external force acting on the atom i and $\vec{u}_i$ satisfies the essential boundary conditions of the problem, $\vec{f}_i \cdot \vec{u}_i$ is the negative work done by the external load acted upon atom i. This is the well-known method of lattice statics. Many of the conventional atomistic formulations (such as the embedded atom method) admit such decomposition although it is not admissible in the more sophisticated density functional approach. For example, the embedded atom method (Daw and Baskes see chapter 2) provides that for a homonuclear material the energy at site i is given by

$$E_i = F_\alpha \left(\sum_{j \neq i} \rho_\alpha(r_{ij}) \right) + \frac{1}{2} \sum_{j \neq i} \phi_{\alpha\beta}(r_{ij}) \qquad (4.4)$$

where F is the embedding energy which is a function of the atomic electron density, ϕ is a pair potential interaction, and α and β are the element types of atoms i and j. The multi-body nature of the EAM potential is a result of the embedding energy term. Both summations in the formula are over all neighbors j of atom i within the cutoff distance. With equation (4.4), the interaction force between atoms can be obtained by calculating the negative gradient of E_i.

Besides EAM potential function, the QC has also been formulated in terms of 3-body interaction potentials of the Stillinger-Weber (SW) type. The main difference between the EAM and SW type of atomistic law is that instead of the embedding energy, F_i, of the EAM, SW includes a term involving three-body interactions to account for the directional bonding in covalent silicon. The energy of an atom i in the SW formulation is then

$$E_i = E_i^{(2)} + E_i^{(3)}$$

$$E_i^{(2)} = \frac{1}{2} \sum_{j \neq i} U_{ij}^{(2)}(r_{ij}) \qquad\qquad (4.5)$$

$$E_i^{(3)} = \frac{1}{6} \sum_{j \neq i} \sum_{k \neq (i,j)} U_{ijk}^{(3)}(r_{ij}, r_{ik})$$

$U_{ij}^{(2)}$ is a pair potential between atom i and its neighbor j and r_{ij} is the interatomic distance. $U_{ijk}^{(3)}$ is the three-body potential and r_{ij} is the vector from atom i to neighbor atom j. In both the EAM and the SW framework, the exact details of the functions ρ_i, F_i, $U_{ij}^{(2)}$ and $U_{ijk}^{(3)}$ are defined to produce a best-fit to various properties of a given material, and thus can be used to describe a wide range of metallic and semi-conducting crystals. In addition to the potential energy of the atoms, there may be energy due to external loads applied to atoms. Thus, the total potential energy of the system (atoms plus external loads) can be written as equation (4.3). It is emphasized by QC developers that the using of atomistic model to determine constitutive relations is not tied to any particular choice of atomistic calculations.

1.3. Continuum Model

It is obvious that the formulation reduces to conventional anisotropic elasticity for small deformations in applications such as MNT and so on. Energy minimization based upon rigid body movement is commonly realized in regions far from lattice defects. The direct computation of the strain energy density using atomistic potential prevent the developing of microscopic structure on a fine scale which can includes lattice defects. Traditional finite element method (FEM) postulates constitutive relation and fit for many engineering applications which cannot give a reasonable explanation about the fundamental mechanism behind the phenomenon. QC differs from classical FEM in that it appeals to the microscopic underpinnings to yield a model for the mechanical response of material. The use of constitutive relations derived in atomistic scale greatly improves the accuracy and the fundamental insight into the crystal deformation and fracture. In the QC computation, the origin in the undeformed configuration tying the continuum finite element model to the underlying crystal structure is defined. The nearest atom is selected as representative atom for quadrature point in the FEM mesh. The immediate crystalline neighborhood in the undeformed configuration surrounding this atom is then deformed according to the continuum deformation field. For local elements the deformation is homogeneous following the local deformation gradient, while for nonlocal elements every atom is displaced according to displacements interpolated to its position resulting in nonhomogeneous deformation. Finite element formulation which serves as the basis of QC includes both local and nonlocal cases. In QC, adaptive finite element method which is well-suited to multiscale analysis is introduced for spatial discretization. The specified region is divided into numerous finite elements while the deformation mapping and the deformation gradients are discretized in manner as following:

$$\phi_h(X,t) = \sum_{i=1}^{n} \phi_i(t) N_i(X) \qquad (4.6)$$

$$F_h(X,t) = \sum_{i=1}^{n} \phi_i(t) \nabla_0 N_i(X) \qquad (4.7)$$

here $i=1,\cdots\cdots,n$ are the nodes in the mesh, n is the number of nodes. $\phi_i(t)$ is the nodal coordinate at time t, $N_i(X)$ is the interpolation or shape function.

1.4. Coupling of Different Physical Model

The basic idea of QC is that the method should reduce to lattice statics in the atomistic limit when the finite element mesh is refined to lattice space. The realization of this condition is based upon the correct coupling of numerical simulation methods at different length scales. This implies the physical quantity such as energy on the atoms computed by different approaches should agree when the QC mesh is fully refined. The efficient solution is to correctly partition the energy for atomistic scheme where the contributions of individual atoms to the total energy can be identified. In QC the energy density related with atomic position which integrated with mesh domain is realized by uniformly distributed of atom energy over the Voronoi polyhedron associated with it and the energy density can be expressed:

$$\varepsilon_i = \frac{e_i}{\Omega} \tag{4.8}$$

here e_i is the energy of atom i, ε_i is the energy density and Ω is the Voronoi volume. Every element in the mesh will experience contribution of energy from Voronoi cells belongs to the atoms at its corners if atom lattice covers continuum finite element mesh with nodes at all atom sites. The total energy of an element can be computed by integrating all atomic energy densities fall within its domain.

$$E_e = \sum_{i=1}^{N} \int_{\Omega_e} W_i(X)dV_i \tag{4.9}$$

here e is the element number, i runs through all atoms in the materials, W_i is the density of atom i equal to ε_i inside Voronoi cell of atom i and zero outside it. The force acted upon atom can be acquired as follows:

$$f_i = \sum_{e=1}^{M} \frac{\partial E_e}{\partial r_i} = \sum_{e=1}^{M} \sum_{j=1}^{N} \int_{\Omega_e} \frac{\partial W_j(x)}{\partial r_i} dV \tag{4.10}$$

where f_i is the force on atom i, r_i is the position vector of atom i relative to its equilibrium position.

As the philosophy of QC is to consider the atomistic description as the "exact" model of material behavior, but at the same time acknowledge that the sheer number of atoms make most problems intractable in a fully atomistic framework. Then, the QC uses continuum assumptions to reduce the degrees of freedom and computational demand without losing atomistic detail in regions where it is required [1]. Recognizing that the nonlocal QC is exactly correct in regions where atomic scale accuracy is needed, while the local QC has the advantage of computational efficiency in regions where the deformation is changing relatively slowly on the atomic scale and the convenience of direct application of continuum boundary conditions, it seems desirable to have the ability to use both formulations in a single simulation. Such a formulation has been developed by combining the local QC and the energy-based formulation of the nonlocal QC.

In the coarse scale-only region, where the molecular dynamics force is unavailable due to the elimination of those atomistic degrees of freedom, an approximation to the external force must be made. The Cauchy-Born rule is a homogenization method by which continuum stress and stiffness measures can be obtained directly from an interatomic potential; the link between the atomistic and continuum is achieved by way of the deformation gradient F. As the Cauchy-Born (CB) rule is a homogenization method, the major assumption invoked when using the approach is that the lattice underlying any continuum point is required to deform homogeneously according to the continuum deformation gradient F. For crystalline solids with a simple lattice structure, this means that every atom in a region subject to a uniform deformation gradient will be energetically equivalent. Thus, the energy within an element can be estimated by computing the energy of one atom in the deformed state and multiplying by the number of atoms in the element. In practice, the calculation of the CB energy is done separately from the model in a "black box", where for a given deformation gradient $\vec{F}$, a unit cell with periodic boundary conditions is deformed appropriately and its energy is computed. The strain energy density in the element is then given by

$$\varepsilon(\vec{F}) = \frac{E_0(\vec{F})}{\Omega_0} \tag{4.11}$$

here Ω_0 is the unit cell volume (in the reference configuration) and E_0 is the energy of the unit cell when its lattice vectors are distorted according to $\vec{F}$. In situations where the deformation is varying slowly from one element to the next and where surface energetics is not important, the local approximation is a reasonably good one. Using the CB rule, the QC can be thought of as a purely continuum formulation, but with a constitutive law that is based on atomistic rather than on an assumed phenomenological form. The CB constitutive law automatically ensures that the correct anisotropic crystal elasticity response will be recovered for small deformations. It is non-linear elastic (again as dictated by the underlying atomistic potentials) for intermediate strains and includes lattice invariance for large deformations.

The local QC formulation successfully imbues the continuum FEM framework with atomistic properties such as nonlinearity, crystal symmetry and lattice invariance. The latter property means that dislocations may exist in the local QC. However, the core structure and energy of these dislocations will only be coarsely represented due to the uniform deformation constraint. The same is true for other defects such as surfaces and interfaces, where the deformation of the crystal is non-uniform over distances shorter than the cut-off radius of the interatomic potentials. While the local QC can support deformations (such as twinning) which may lead to microstructures containing such interfaces, it will not account for the energy cost of the interface itself. In order to correctly capture these details, the QC must be made nonlocal. The nonlocal QC has been formulated in two ways, namely, the energy-based formulation and the force-based formulation. In practice, this involves computing the total energy and the forces (derivatives of the total energy) on each repatom, and then driving the system to the energy minimizing (zero force) configuration. In the energy-based formulation, the energy of each repatom is computed from the deformed neighbor environment dictated from the current interpolated displacements in the elements. The forces on all the repatoms can be obtained as the derivatives with respect to the repatom positions and energy minimization can proceed. In the force-based QC formulation, the starting point is to recognize that energy minimization physically corresponds to solving for the configuration for which the force on each degree of freedom is zero. Equilibrium can be sought be working directly from an approximate expression for the forces, rather than from the explicit differentiation of an energy functional.

The fully non-local description has certain disadvantages, the main one being the significant increase in computational cost as compared with the local approach. Each energy or force evaluation requires the mapping of a cluster of atoms and their neighbors to their deformed configuration at every repatom, followed by the necessary interatomic potential evaluations necessary to compute the energy or forces for the cluster. This is more than is required in the local calculation, but can be made reasonably efficient with careful bookkeeping and look-up tables. While the local QC suffered from an inability to resolve any energy of free surfaces or interfaces, the nonlocal QC suffers from a significant overestimate of surface effects. Consider a repatom at the corner of a cubic specimen. The cluster for this repatom will be a high energy cluster due to the three free surfaces it will see. At the same time, if this repatom has a large weight because it represents a large volume of material, the resulting energy will be as though that entire volume of material is located in the vicinity of a corner. The result is a significant overestimate of the energetic importance of the corner and therefore spurious relaxations of this repatom. The main advantage of the nonlocal QC, both energy and force-based, is that when it is refined down to the atomic scale, it reduces exactly to lattice statics, correctly capturing details of dislocation cores, stacking faults and grain boundaries.

Because the nonlocal QC is exactly correct in regions where atomic scale accuracy is needed, while the local QC has the advantage of computational efficiency in regions where the deformation is changing relatively slowly on the atomic scale and the convenience of direct application of continuum boundary conditions, it seems desirable to have the ability to use both formulations in a single simulation by combining the local QC and the energy-based formulation of the nonlocal QC as presented above. As in the energy-based nonlocal QC, the energy can be approximated by computing only the energy of the repatoms, but this time assume that we can treat each repatom as being either local or nonlocal depending on its deformation environment. Thus, the repatoms are divided into local repatoms (N_{loc}) and nonlocal repatoms (N_{nl}) and $N_{loc} + N_{nl} = N_{rep}$. The energy expression is then approximated as

$$E^{tot,h} \approx \sum_{i=1}^{N_{nl}} n_i E_i(\vec{u}_h) + \sum_{i=1}^{N_{loc}} n_i E_i(\vec{u}_h) \tag{4.12}$$

The weight n_i for each repatom (local or nonlocal) is determined from a tessellation that divides the body into cells around each repatom. One physically sensible tessellation is Voronoi cells. When making use of the mixed formulation described in equation (4.12), it now becomes necessary to decide whether a given repatom should be local or nonlocal. This is achieved automatically in the QC using nonlocality criterion. The key feature that should trigger a nonlocal treatment of a repatom is a significant variation in the deformation gradient on the atomic scale in the repatom's proximity. Thus, the nonlocal criterion in implemented as follows. A cutoff is defined r_{nl}, empirically chosen to be between two and three times the cutoff radius of the interatomic potentials. The deformation gradients in every element within this cutoff of a given representative atom are compared, by looking at the differences between their eigenvalues. The criterion is then:

$$\max_{a,b;k} \left| \lambda_k^a - \lambda_k^b \right| < \varepsilon \qquad\qquad (4.13)$$

where λ_k^a is the kth eigenvalue of the right stretch tensor $\vec{U}_a = \sqrt{\vec{F}_a^T \vec{F}_a}$ in element a, $k = 1 \cdots 3$, and the indices a and b run over all elements within r_{nl} of a given repatom. The repatom will be made local if this inequality is satisfied, and nonlocal otherwise. In practice, the tolerance ε is determined empirically. A value of 0.1 has been used in a number of tests and found to give reasonable results.

2. MULTISCALE FINITE ELEMENT METHODS (MsFEM)

Although small-scale features of the materials are important, the large-scale connectivity can play a crucial role. When both fine- and coarse-scale information are combined, the resulting materials properties have scale disparity and vary over many scales. As the physical processes become complicated due to additional physics arising in multiphase flows in many machining techniques such as CMP, it becomes impossible to simulate these processes without coarsening model equations. When performing simulations on a coarse grid, it is important to preserve important multiscale features of physical processes. Multiscale methods compute effective properties of the media in the form of basis functions which are used, as in classical upscaling methods, to solve the processes on the coarse grid, the materials properties often contain uncertainties. These uncertainties are usually parameterized and one deals with a large set of permeability fields (realizations) with a multiscale nature. This brings an additional challenge to the fine-scale simulations and necessitates the use of coarse-scale models. The multiscale methods are important for such problems. For these problems, one can look for multiscale basis functions that contain both spatio-temporal scale information and the uncertainties. These basis functions allow us to reduce the dimension of the problem and simulate realistic stochastic processes. The multiscale finite element methods (MsFEM) which put up by Y. Efendiev and T. Y. Hou [5] can easily be generalized to take into account both multiscale features of the solution and the associated uncertainties. Multiscale finite element methods (MsFEMs) share similarities with upscaling methods which have been commonly applied and are effective in many cases. The main idea of upscaling techniques is to form coarse-scale equations with a prescribed analytical form that may differ from the underlying fine-scale equations. In multiscale methods, the fine-scale information is carried throughout the simulation and the coarse-scale equations are generally not expressed analytically but rather formed and solved numerically. For problems with scale separation, one can establish the equivalence between upscaling and multiscale methods. Furthermore, MsFEMs also share similarities with variational multiscale methods. In this approach, the solution of the multiscale problem is divided into resolved (coarse) and unresolved parts. The objective is to compute the resolved part *via* the unresolved part of the solution and then approximate the unresolved part of the solution. This is shown in the framework of linear equations. Typically, the approximation of the unresolved part of the solution requires some type of localization. The localization leads to methods similar to the MsFEM.

Another important feature of MsFEMs is the use of variational formulation at the coarse scale which makes it possible to couple multiscale basis functions. Fine-scale formulation of the problem that allows computing multiscale basis functions is not necessarily based on partial differential equations and can have a discrete formulation. In this regard, MsFEMs share conceptual similarities with some approaches that couple atomistic (discrete) and continuum effects. These approaches use a variational formulation at the coarse scale, but use a discrete atomistic description at finest scales (*e.g.*, quasi-continuum method). Here, the notion "multiscale finite element methods" refers to a number of methods, such as multiscale finite volume, mixed multiscale finite element method and so on. The concept that unifies these methods is the coupling of oscillatory basis functions *via* various variational formulations. The central goal of MsFEM is to obtain the large-scale solutions accurately and efficiently without resolving the small-scale details. The main idea is to construct finite element basis functions that capture the small-scale information within each element. The small-scale information is then brought to the large scales through the coupling of the global stiffness matrix. The basis functions are constructed from the leading-order homogeneous elliptic equation in each element. As a consequence, the basis functions are adapted to the local microstructure of the differential operator. MsFEMs consist of two major ingredients: multiscale basis functions and a global numerical formulation that couples these multiscale basis functions. Basis functions are designed to capture the multiscale features of the solution. Important multiscale features of the solution are incorporated into these localized basis functions which contain information about the scales that are smaller (as well as larger) than the local numerical scale defined by the basis functions. A global formulation couples these basis functions to provide an accurate approximation of the solution.

The general framework of MsFEM can be expressed by means of an example as follows:

$$Lp = f \tag{4.14}$$

here L is an operator and the p can be approximate on coarse grid using MsFEM. If W_h is assumed as an assemble of finite dimensional space while h is a scale of computation (coarse grid). Then the multiscale basis functions are replaced by multiscale maps defined as

$$E^{MsFEM} : W_h \rightarrow P_h \tag{4.15}$$

while for each element $v_h \in W_h, v_{r,h} = E^{MsFEM} v_h$ is defined as

$$L^{map} v_{r,h} = 0 \quad in \ K \tag{4.16}$$

here L^{map} can be different from L, v_h denotes the coarse scale approximation and $v_{r,h}$ denotes the fine scale approximation. The L^{map} enable the ability of study the effects of small scales. Moreover, the domains different from the target coarse block K can be used in the computations of the local solutions. To solve (4.16) one needs to impose boundary and initial conditions. This issue needs to be resolved on a case by case basis, and the main idea is to interpolate v_h onto the underlying fine grid. To find a solution of (4.14) in P_h, one can substitute P_h (which denotes a coarse-scale solution defined in W_h) into (4.14) discretized on the fine grid. Because P_h is defined on the coarse grid, the resulting system is projected onto the coarse-dimensional space. This can be done in various ways. A common approach is to multiply the resulting fine-scale system by coarse-scale test functions; that is find $P_h \in W_h$ (consequently $p_{r,h} \in P_h$) such that

$$\left\langle L^{global} p_{r,h}, v_h \right\rangle = \left\langle f, v_h \right\rangle, \quad \forall v_h \in W_h \tag{4.17}$$

here $\langle \bullet, \bullet \rangle$ denotes the duality between X and Y. The fine scale system $L^{global} p_{r,h}$-f can be realized by multiplying coarse scale test functions. The main feature of MsFEMs is the use of a variational formulation at the coarse scale that makes it possible to couple multiscale basis functions. Multiscale basis functions or multiscale maps defined by (4.16) are not necessarily based on partial differential equations and can have a discrete structure and satisfy a discrete equation at the fine grid. It is evident from the above abstract formulation that L^{map} is used only for the computation of $p_{r,h}$ (given p_h) and the variational formulation (4.17) can be chosen in different ways depending on the problem. One can consider general applications of MsFEMs involving discrete problems where the basis functions satisfy discrete systems. For example, one can consider an application where the coarse-scale equations have a continuum formulation and describe porous media flows, whereas the local problems are discrete and solved *via* the pore network model.

3. BRIDGING DISCRETE TO CONTINUUM SCALE

The continuum formulation in the mixed atomistic-continuum model is retained as the overall framework while the atomistic input is introduced by way of constitutive description of material. The constitutive relation serves as the gateway for bringing atomistic process into the continuum realm enable the mixed model to span several length scales which is benefit of study the mechanism in depth. Calculation of macroscopic properties of materials by coupling model of different scales is meaningful on understanding fundamental essence of materials deformation behavior in MNT.

New classification of multiscale methods is proposed based on how the atoms in the atomistic domain are connected to the continuum domain. In this classification, approaches that relate atoms and finite element nodes in a one-to-one manner, or through a form of interpolation, will be referred to as direct coupling (DC) methods. According to this definition, most of the existing multiscale methods belong to DC methods. While DC methods are straightforward, there is a difficult for truly seamless transition along the boundaries between the atomistic and the continuum domain. This is caused by the intrinsic incompatibility between materials at the two sides of the interface (or boundary) which causes nonphysical phenomena, including the so-called ghost forces. This incompatibility arises because the constitutive behavior of the continuum on one side is local in nature but that of atoms on the other side is non-local. Here, non-local constitutive behavior indicates that the force at any atom depends also on atoms which do not directly connect with the atom but operate in its neighborhood through inter-atomic forces; local behavior indicates that the force (stress) of a material point depends only on the deformation gradient (strain) at the same point of the continuum. Due to this local behavior, atoms in the continuum region cannot feel interactions from other atoms nearby as their counterparts in the atomistic region can. The other shortcoming of DC methods is that in most cases the finite element size has to be nearly the inter-atomistic distance at the atomistic/continuum interface to perform well. This feature causes substantial difficulty when the model sizes are increased.

As DC methods need to have an extremely small mesh size at the interface of FEM domain, they will have substantial difficulty in meeting this new challenge. An alternative to DC methods called ESCM (embedded statistical coupling method) is proposed recently using statistical averaging over selected time interval and volume in atomistic subdomains at the MD/FE interface to determine nodal displacement for the continuum FE domain. Another non-DC method called GP (generalized particle dynamics) method is proposed using constant material neighbor link cells at the interface region to mutually transfer information from bottom-scale up or from top-scale down to quantitatively link variables at different scales. Because the GP method conducts calculation of all scales of generalized particle in each corresponding atomistic scale with the same potential function and numerical algorithm as the atomistic scale, this method is also called extended molecular dynamics.

3.1. Coupling FE & Atomistic Method Without Handshake Region

Recently, quasi-static coupling technique was used in the development of the CADD (coupled atomistic and discrete dislocation) method [6]. In this approach, there is no handshake region and strong compatibility is enforced. The focus of this development was the connection to discrete dislocation methods in the continuum region. Incorporating elastic dislocations necessitates the use of linear elasticity (as opposed to the Cauchy-Born rule) in the continuum region, but in the limit where there are no dislocations in the continuum, CADD can use the Cauchy-Born rule or any other nonlinear constitutive law in the finite elements. In this limit (no dislocations and the Cauchy-Born rule) CADD shares the features of the QC method, but with a force-based rather than energy-based coupling scheme. In fact, CADD can be described as a force-based QC formulation. The advantages in doing this is that defects, mainly dislocations, generated within the atomistic region are allowed to pass through the atomistic/continuum border into the continuum, where they are represented *via* discrete dislocation mechanics. Because discrete dislocation mechanics incorporate the elastic stress field emitted from a dislocation into the continuum stress and modulus expressions, the defects are able to be tracked once they pass into the continuum region. In CADD, there is no handshake region and strong compatibility sets the positions of the padding atoms and the nodes along the heavy. The forces on every atom at interface are computed as if the continuum did not exist, from the derivative with respect to atom positions of energy functional.

The total energy of the coupled system is

$$E(\vec{u}, \vec{d}^M) = E^C(\vec{u}) + E^M(\vec{d}^M) - E^{ext} \tag{4.18}$$

here E^M and E^C are the energies of molecular and continuum domains respectively, E^{ext} is the work of external forces. The vector $\vec{u}$ is the displacement field in the continuum and is a vector of discrete molecular displacements. The continuum energy is given by

$$E^C(\vec{u}) = \int_{\Omega_0^C} \omega_C(\vec{F}) d\Omega_0^C \tag{4.19}$$

where ω_C is the strain energy density, $\vec{F}$ is the deformation gradient and can be expressed as

$$\vec{F} = \vec{I} + \frac{\partial \vec{u}}{\partial \vec{X}} \tag{4.20}$$

The displacement field can solved if finite element method is used for continuum

$$\vec{u}(\vec{X}) = \sum_{I \in \Omega} N_I(\vec{X}) \vec{u}_I \tag{4.21}$$

here $N_I(\vec{X})$ is the shape function expressed as the function of nodal coordinate $\vec{X}$, $\vec{u}_I$ is the nodal displacement, Ω is the set of all FEM nodes.

The potential energy of the bond *I-J* is denoted by $\omega_{IJ}^M = \omega_M(X_I, X_J)$ for a two-body potential, where X_I is the current position of atom *I*. The atomistic and displacements at the interface ought to be taken to be equal. Define common vector $\vec{d} = \{d_I\}_{I=1}^n$, where n is the number of independent nodes and atoms and the compatibility is enforced by letting

$$d_I = \begin{cases} d_I^M & X_I \in \Omega_0^M \\ d_I^M = u_I & X_I \in \Omega_0^M \cap \Omega_0^M \\ u_I & X_I \in \Omega_0^C \end{cases} \tag{4.22}$$

The energy of the model can then be written as

$$E(d) = E^C(d) + E^M(d) - E^{ext} \tag{4.23}$$

Equilibrium configurations corresponding to the stationary points may lead to the following results:

$$\frac{\partial E^M}{\partial d_I} + \frac{\partial E^C}{\partial d_I} - \frac{\partial E^{ext}}{\partial d_I} = 0 \tag{4.24}$$

These derivatives corresponding to nodal forces and atomistic forces which can be written as:

$$f_I^M + f_I^C - f_I^{ext} = 0 \qquad (4.25)$$

The positions of atoms which lies in inside Ω_0^C can be obtained by the finite element interpolation

$$d_J^M = \sum_{I \in S} N_I(X_J) u_I \qquad (4.26)$$

The governing equations are identical to those described for coincident coupling, the force on the atoms can be added to continuum nodes.

3.2. Bridging Domains with Handshake Region

One way to mitigate the unphysical phenomena, such as the artificial heating in the MD region by wave reflection and the ghost force at the interface, is to make the transition from atomistic region to continuum less abrupt by using a handshake region. The desired smooth transition is somewhat ad hoc; it can be achieved, however, by the simple blending of energy in the handshake region. Here, the bridging domain is the handshake region [6-7]. The bridging domain method is in essence an overlapping domain decomposition scheme where the compatibility in the overlapping (handshake) domain is enforced by Lagrange multipliers. The advantage of this method is that the atomic nuclei need not be coincident with the nodes of continuum mesh. Furthermore, the method is generally used with a linear scaling of the energies in the handshake domain, which enables the methods to alleviate the errors that arise from dropping atomistic energies due to far field atoms and provides a gradual transition from the molecular model to the continuum model.

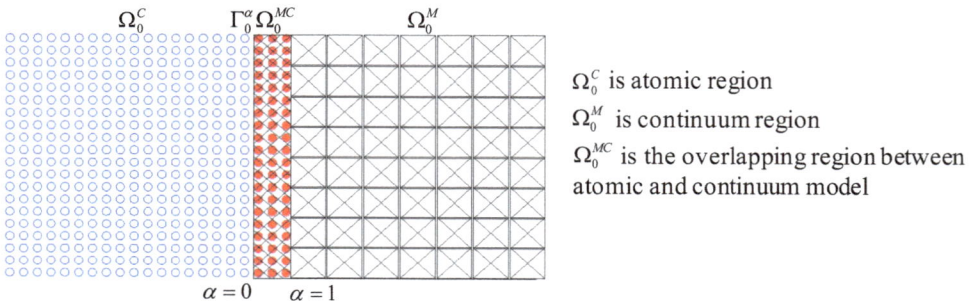

Ω_0^C is atomic region

Ω_0^M is continuum region

Ω_0^{MC} is the overlapping region between atomic and continuum model

Figure 1: Coupling atomistic and continuum model with handshake region

For energy-based method with handshake region, the total energy should include the Hamiltonian in the atomistic, continuum and handshake region and can be given by:

$$E^{total} = E_\alpha^{Continuum} + E_\alpha^{Atom}$$

$$E_\alpha^{Continuum} = \int_{\Omega_0^C} \alpha(\vec{X})\omega_C(\vec{F})d\Omega_0^C \qquad \text{(continuum region)}$$

$$E_\alpha^{Atom} = \frac{1}{2}\sum_{I,J\in M}(1-\alpha_{IJ})\omega_{IJ}^{Atom} \qquad \text{(atomic region)}$$

$$\textbf{(4.27)}$$

Here $\alpha_{IJ} = \frac{1}{2}(\alpha_I + \alpha_J)$, $\alpha_I = \alpha(X_I)$, $\alpha(X)$ is energy scaling function in the overlapping subdomain and defined as

$$\alpha(X) = \begin{cases} 0 & \text{in } \Omega_0^M \setminus \Omega^{MC} \\ l(X)/l_0(X) & \text{in } \Omega_0^{MC} \\ 1 & \text{in } \Omega_0^C \setminus \Omega_0^{MC} \end{cases} \qquad \textbf{(4.28)}$$

where $l(X)$ is the least square projection of X onto Γ_0^α while $l_0(X)$ is defined by the distance between the overlapping domain boundaries Γ_0^α on the orthogonal projection of X. The scaling parameters vanish at the edge of the continuum and is unity at the other edge of Ω_0^{MC}.

The displacement of compatibility in the handshaking region is enforced by the constraint:

$$g_I = \|u(X_I) - d_I\|^2 = \sum_{i=1}^{dim}[u_i(X_I) - d_{iI}]^2 = 0 \qquad \textbf{(4.29)}$$

here dim is the dimension. A single constraint which makes the transition from atomistic region to continuum less abrupt is applied on each atom in the transition area and the constraint is realized by Lagrange multiplier.

The Lagrange multiplier can be acquired by finding the stationary point of the following equation:

$$W = E_\alpha^C + E_\alpha^M - E_\alpha^{ext} + \vec{\lambda}^T \vec{g} \tag{4.30}$$

here $\vec{\lambda} = \{\lambda_{il}\}$ is the vector of Lagrange multiplier and $\vec{g} = \{g_{il}\}$. The Lagrange multiplier is generalized force which enforces the constraint g_{il}. The Lagrange multiplier field is expressed in terms of shape function:

$$\vec{\lambda}(\vec{X},t) = \sum_{I \in S^\lambda} \vec{N}_I^\lambda(\vec{X})\overline{\lambda}_I(t) \tag{4.31}$$

here S^λ is the set of all nodes of the Lagrange multiplier mesh, $\overline{\lambda}_I$ denotes Lagrange multiplier at the Lagrange multiplier nodes. Generally, the shape functions for the Lagrange multiplier field will differ from those for the displacements. The Lagrange multiplier field is usually constructed by auxiliary finite element triangulation of the intersection domain. In reality, the bridging domain method works beat as the Lagrange multiplier field is Dirac delta functions at the locations of nuclei. The choice of Lagrange multiplier enforces compatibility exactly with the finite element approximation for the continuum. This approximation for the Lagrange multiplier can be expressed as:

$$\lambda(X) = \sum_{I \in M^{MC}} \hat{\lambda}_I \delta(X - X_I) \tag{4.32}$$

$\hat{\lambda}_I$ is the unknown values of Lagrange multiplier and the $\delta(\bullet)$ is the Dirac delta function.

4. MULTISCALE MODEL ABOUT MNT

4.1. Multiscale Simulation of Microcutting Flat Surface

Figs. (**2 and 3**) give the simulation results of MNT acquired by multiscale method. In this case, the common package QC is adopted for this investigation and the simulation is carried out in two dimension space. The cell dimension along the out-of-plane direction, dz=2.85111, is the same for the two grains we have defined. Note that for many boundaries, the two grains will have incommensurate periodicities along the out-of-plane direction. The QC will correctly treat grains with different values of dz, but only insofar as it will correctly compute the energy and forces on the structure. It is not capable of relaxing the structure in the out-of-plane direction, which generally would require accommodations such as misfit dislocations with *x*- or *y*- line directions and other deformations where the displacement field varies along the *z* direction. Other important parameters used in this study are listed in Table **1**. Multiscale simulation technique enables people's capability of studying physical essence of engineering applications with multiple scale features such as MNT. Fig. (**2**) gives the ultraprecision machining model with micrometer dimension and nanometer level depth of cut. The larger amount of atoms generated by molecular dynamics simulation method makes it unfit for MNT research. The atomic region adapts and expands, as dislocations gliding incline to the *x* direction with about 45° (shear angle) are nucleated and extend into the bulk material (Fig. (**3**)). Materials removal is realized by means of atomic cluster at the initial stage and also the plastic deformation of cutting tool is observed.

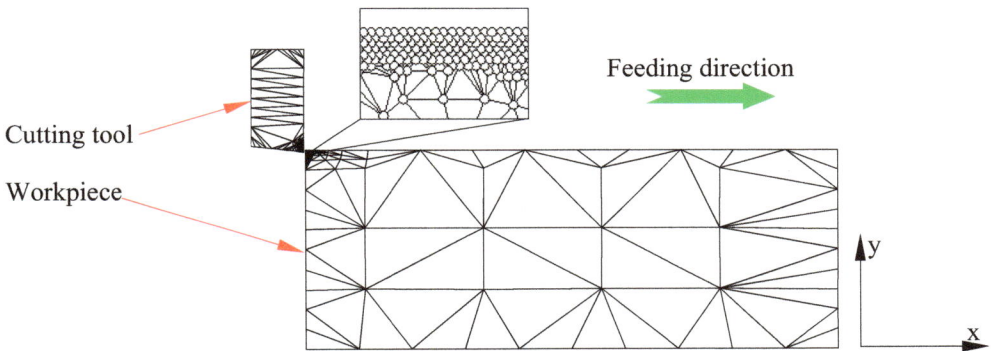

Figure 2: Multiscale model of MNT.

Table 1: Computation parameters.

Workpiece materials	Aluminum
Potential function	EAM
Geometry dimension	$2\mu m \times 1\mu m$
Rake angle	0°
Clearance angle	8°
Cutting speed	1m/s
Depth of cut	10nm
Timestep	0.2×10^{-15}s

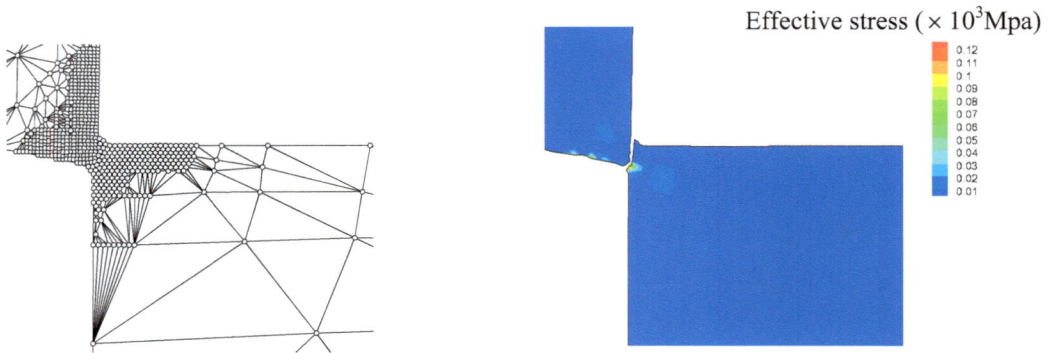

(*a*) Timestep: 70

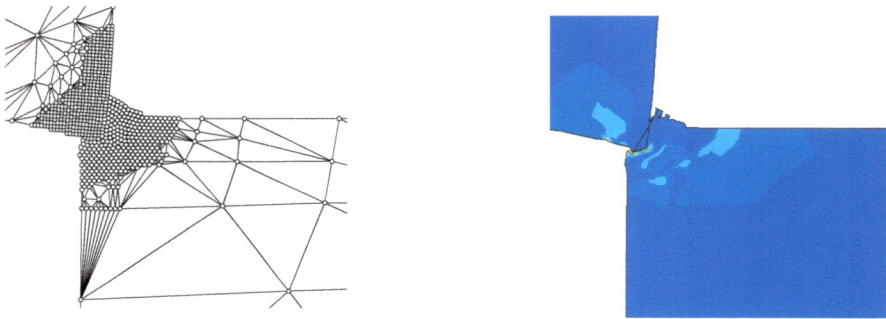

(b) Timestep:210

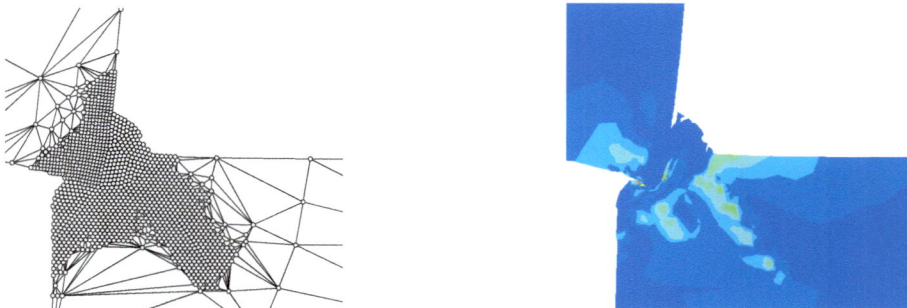

(c) Timestep:350

Figure 3: Multiscale modeling of MNT.

[1] Black-Initial position of tool tip

[2] Magenta to Green-Deformed position of tool tip

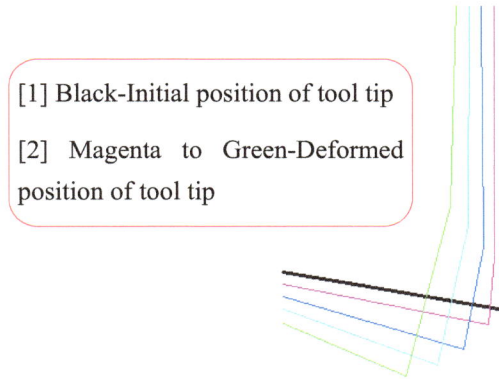

Figure 4: The evolution of tool tip in MNT.

Fig. (**4**) gives the evolution of structural configuration of cutting tool tip. The local rake angle of cutting tool gradually decreases while the clearance angle gradually increases. Small rake angle will improve the strength of cutting tool and the planarization of the machined surface. The local deformation of the tool tip may change the depth of cut which will influence the final surface quality.

4.2. Multiscale Modeling of Microcutting Rough Surface

Fig. (**5** and **6**) gives the multiscale simulation results of microcutting rough surface while the corresponding computation parameters are shown in Table **2**. In reality, there are numerous rough peaks randomly distributed on the engineering surface which makes the many materials removal process happened between the cutting tool and the surface peak. There is several reasons account for this, the mall depth of cut, the wear of cutting tool or the deformation of workpiece. The simulation results justify that different mechanism exist in this process. First, all of the deformations are generated above the surface with no dislocation extending into the bulk materials. Second, most of the surface peak is removed by means of shearing. As the discrete distribution of surface peak which is different from continuum bulks, the periodic impact will further decrease tool life.

Multiscale modeling of MNT under two different technique conditions is carried out which justifies the feasibility of multiscale modeling. The transition of effective stress between atomic region and continuum region is smooth and reasonable which justifies the reasonable of the physical model. Many areas of

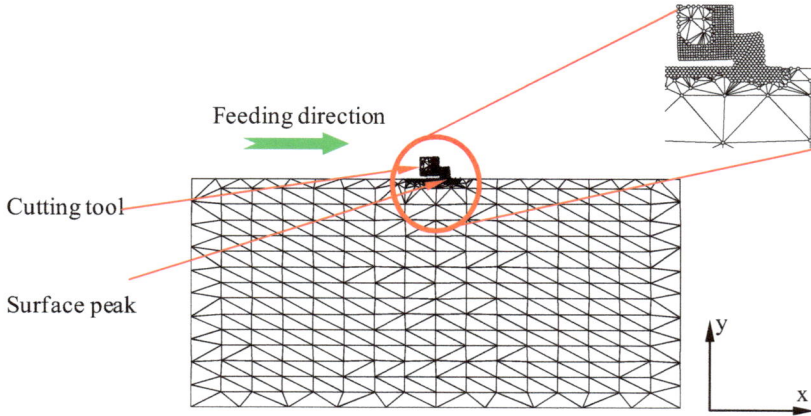

Figure 5: Multiscale model of cutting single peak.

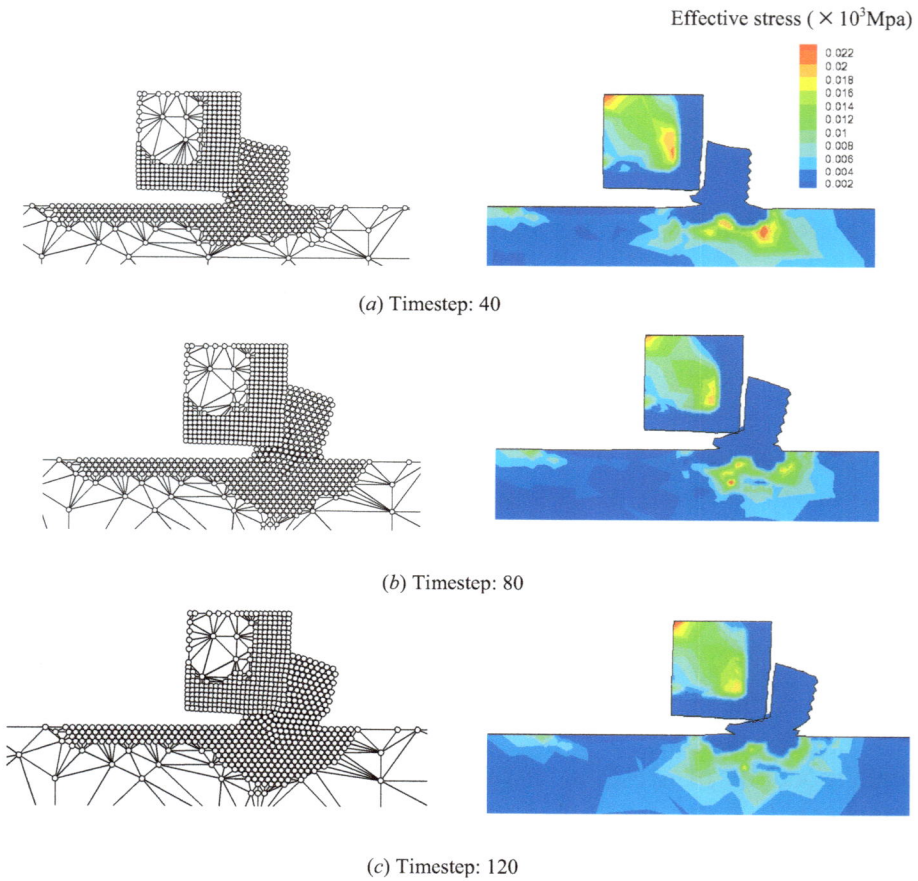

Effective stress (× 10³Mpa)

(*a*) Timestep: 40

(*b*) Timestep: 80

(*c*) Timestep: 120

Figure 6: Multiscale modeling of cutting rough surface.

Table 2: Computation parameters.

Workpiece materials	Aluminum
Potential function	EAM
Geometry dimension	$1\mu m \times 0.5\mu m$
Rake angle	$0°$
Clearance angle	$0°$
Cutting speed	$1m/s$
Depth of cut	$15nm$
Timestep	$0.2 \times 10^{-15}s$

research are rapidly advancing owing to the combined efforts of science and engineering. Owing to the combination of constantly increasing computational power and the increased knowledge and understanding of material behavior, multiscale modeling methods will gradually emerge as the tool of choice to link the mechanical behavior of materials from the smallest scale of atoms to the largest scale of structures.

5. CURRENT STATUS AND CHALLENGES

Comparing with traditional machining technique, MNT is a novel area which includes many new features and challenges. For instance, in line with the rapid progress in miniaturization and high integration in electronic components, the dimensional accuracy of the components is shifting from on the order of micrometers ($10^{-6}m$) to nanometers ($10^{-9}m$). The manufacturing precision has already been improved from micrometer level to nanometer level. It has already become the goals of all high technologies to acquire super-smooth and defect-free surface. Such requirement has approached the limit of manufacturing technology (Extreme Manufacturing). Under this circumstance, any change of the flow field or conflict in physical-chemical factors could deteriorate the surface quality, any tiny hard particle may generate large pits or mark in the surface. Further study on physical aspects of extreme manufacturing technology is crucial for obtaining planarization surface with nanometer level roughness.

Despite intense theoretical and experimental research on MNT, there is still serious lack of fundamental understanding of this process. Although MNT has already been accepted in industry, its application still rests on the semi-empirical stage. It is different from traditional machining technique as the depth of cut or materials removal volume in MNT is minimal which makes it impossible to extract effective data in the experimental results. Presently, the researchers can't give convinced illumination about the mechanism of MNT. The reason for this is that the MNT process is a complex system characterized by multiphase, multiscale and multilevel process. It is not the geometrical downsizing but should exist some new discipline dominates the material remove and surface generation process in the MNT. Real materials are highly complex involving structure on many different length scales. The observed macroscopic behavior is often a function of collective microstructure of the material with no one length scale able to be singled out as dominant. Multiscale methods offer the best hope for bridging the traditional gap that exists between engineering application and the theoretical approach for studying and understanding the physical essence of MNT.

Multiscale method differs from classical finite element simulation as the constitutive input into the formulation is derived from explicit atomistic simulation results based on interatomic potentials rather than phenomenological models. The bridging length scales is realized by recourse mesh refinement and adaptive meshing techniques. The automatic seamless transition between the microscopic region and macroscopic region is acquired by the mesh down to the lattice scale near the an harmonic region at the same time retaining coarse descriptions in less deformed region away from machining tool. The use of real atomic structure to compute the constitutive response of a point in the continuum is that crystalline structures normally transparent to continuum formulations can be modeled in a straightforward fashion. Near defect cores almost all atoms are sampled with the effect that the method essentially reduces to an atomistic lattice statics computation while in regions behaving elastically very few atoms are sampled with each one representing large numbers of its neighbor that are behaving essentially the same. The numbers of degrees of freedom need to be resolved are greatly reduced without sacrificing the atomic resolution.

Presently, there are still many of the fundamental features of multiscale model needs to be further studied and clarified to make multiscale method a meaningful quantitative tool for engineering applications. The mesh dependence of the results needs to be quantified and the mesh coarsening must be introduced into the model. Currently, the mesh adaption limited to gradually refine the mesh as the deformation progresses. However, if the mesh is not allowed to be coarsen as defects move out of one region and into another, the final result will be that the mesh will become fully refined everywhere and the multiscale method will degenerate into atomistic method. Thermal coupling between the atomistic and continuum regions is necessary as most materials are used at a temperature higher than zero absolute temperature. For problems involving significant temperature changes, such as micro-machining, thermal coupling has to be addressed properly in order to obtain valid results. The multiscale simulation algorithms can be extended to include other scales, such as the tight binding (TB). Dynamic multiscale methods depend on continuum mechanics and thermodynamics for the description of the continuum regions, while the atomistic region is of course described by an empirical potential within the framework of molecular dynamics. Methods to make direct links between the atomic scale and the macroscale, such as the Cauchy-Born rule and the constitutive laws derived therefrom, are the natural choice to define the response of the continuum region and ensure a good coupling to the atomistic region. Clearly, statistical mechanics plays an important role in developing the approximations for the free energy expressions, and they will no doubt play a future role as these methods are extended to correctly describe heat flux between the atomistic and continuum domains in the nonequilibrium methods.

REFERENCES

[1]   E.B. Tadmor, "The quasicontinuum method", PhD Thesis, Brown University, Rhode Island, USA, 1996.

[2]   R.E. Miller, E.B. Tadmor, R. Phillips, M. Ortiz, "Quasicontinuum simulation of fracture at the atomic scale", *Model. Simul. Materi. Sci. Engine.*, vol. 6(5), pp. 607-638, June 1998.

[3]   J. Knap, M. Ortiz, "An analysis of the quasicontinuum method", *J. Mecha. Phys. Solids*, vol. 49(9), pp. 1899-1923, April 2001.

[4]   R.E. Miller, E.B. Tadmor, The quasicontinuum method: Overview, applications and current directions, *J. Compu. Aided Materi. Desi.* Vol. 9, pp. 203-239, September 2003.

[5]   Y. Efendiev, T.Y. Hou, *Multiscale Finite Element Methods*. New York: Springer, 2009.

[6]   J. Fish, *Multiscale methods-bridging the scales in science and engineering*. Oxford: Oxford University Press, 2009.

[7]   J.H. Fan, Multiscale analysis of deformation and failure of materials. New York: John Wiley & Sons, 2011

Send Orders for Reprints to reprints@benthamscience.net
Micro and Nanomachining Technology Size, Model and Complex Mechanism, 2014, 221-270   221

CHAPTER 5

Complexity of MNT

Abstract: The physical process involved in the MNT is unique which makes it a special area of machining science. The reason for this is that the minimizing of the workpiece or machining precision level is greatly improved in the MNT comparing with the traditional machining technique. Mechanical properties of materials are scale dependent based on the strain gradient plasticity and the effect of dislocation-assisted sliding. When material is removed by machining, there is a substantial increase in the specific energy required with decrease in chip size. It is generally believed that this is because all metals contain defects (grain boundaries, missing and impure atoms, *etc.*), and, when the size of the material removed decreases, the probability of encountering a stress-reducing defect decreases. Furthermore, the dominate role of the volume force in the larger scale is replaced by the surface force such as adhesive force at this micro & nano-length scale. The generation and dissipation of heat and the materials plastic deformation also involved different control mechanism from the macroscopic scale. Traditional metal cutting principle cannot give a reasonable explanation about phenomenon in the MNT. MNT is a novel challenge that should be studied in depth using ideas of modern physics that deals with the problems of complexity. This chapter devoted to the introduction of studying MNT using complexity theory (self-organization, nonequilibrium thermodynamics, fractal theory *etc.*) and some new corresponding developments.

Keywords: Complexity, dynamic evolution, self organization, dynamic equilibrium, energy dissipative mechanism.

INTRODUCTION

Feynman, the Nobel Laureate in physics, expatiated the scaling down of lathes and drilling machines, and talks about drilling holes, turning, molding, stamping parts, and so forth. Feynman also described the need for micro- and nanomanufacturing as the basis for creating a microscopic world that would benefit mankind. MNT is a challenge presented to us to produce single-nanoscale artifacts in a mass production fashion that obviously produces the accompanying economies of scale. MNT is a novel area which is not only the improvement of the precision of machining and the measurement but also the ability of people's understanding and transforming has been improved from macroscopic scale to

Xuesong Han (Ed)

microscopic scale. The control mechanism of the MNT is not the similarity reduction but should be some new mechanism plays a dominant role. It is necessary to redefine some macroscopic physical property such as elastic modulus, density and temperature. The common principle such as *O Jamie de geometry*, *Newton mechanics*, *Macroscopic Thermodynamics* and *Electromagnetics* are not applicable to phenomenon happened in nanometer scale while the quantum effects, wave characteristics and the microscopic fluctuation become the dominant factor. Machining is a complex phenomenon associated with a variety of different mechanical, physical, and chemical processes. Traditionally, this area of applied science and engineering was considered primarily a mechanical process. However, recently, MNT has been viewed as a more complex phenomenon. There are many factors with complicated interactions involved in the MNT which makes it a systematic engineering.

1. DYNAMIC EVOLUTION AND SELF-ORGANIZATION OF MNT

The technical function of MNT system heavily depends on the force and motion. According to its basic physical definition, the term motion denotes the change of the position of an object with time. In a common sense, the study of how things change with time and of the forces that cause them to do so is the objective of dynamic investigation. To perform machining operations, relative motion is required between the tool and workpiece. This relative motion achieved in most machining operation is a combined motion consisting of several elementary motions as the primary motion, called the cutting speed and the second motion called the feeding feed. The tool geometry and tool setting relative to the workpiece, combined with these motions, produce the desired shape of the machined surface. While performing the basic motions to shape the workpiece, the tool requires energy which is normally represented by the cutting forces and velocities in the corresponding directions. This energy determines tool life, integrity of the machined surface and accuracy of machining. Therefore, a simple and reliable method to determine this energy has to be developed to improve cutting tools and the machining operation as the whole.

Traditional turning is a general term for a group of machining operations in which the workpiece carries out the prime rotary motion while the tool performs feed motion whose main function is to remove materials. These are used for the external and internal turning of surfaces. The workpiece is clamped in a self-centering three jaw lathe chuck installed on the machine spindle that provides rotation and the tool installed on the tool post which is a part of a lathe carriage that provides the feed motion. Furthermore, the depth of cut gradually decreased with the improving of machining precision. Under this circumstance, most of the work is completed by the tool tip which means that any tiny mechanical vibration or abrupt physical-chemical variation will result in the surface damage or decreasing of the precision. Therefore, MNT is a complex engineering include many high technique such as machine tool design, nanometer metrology, mechanical error control, thermal drift control and so on. The evolvement of MNT accompanies the variation of the relative motion and the constraint mode, namely, the alternate of force and motion. Take the example of ceramics machining, pure precision single point cutting often generate crack or pits in the surface which lead to the development of multiple point machining-precision grinding. Comparing with precision cutting, a more complex relative movement track is formed in the precision grinding which improve the machining quality. Furthermore, there is still rigid constraints (fixed grain) about the micro-cutting tool which result small surface defects. The driver for higher quality machined surface induced the development of no constraint (free grain) multiple body machining technique-lapping or polishing. This two machining techniques is by far the highest precision machining method which benefit from the novel relative motion and the external constrains. In this sense, the enhancement of the MNT is the direct result of dynamic evolution of the physical system.

MNT, in general, is also an irreversible process because energy dissipation takes place. This process changes and develops its characteristics and parameters with time, similar to that seen in evolution. Therefore, we should be interested not only in the initial and final stage of the definite MNT but also in the way this process has developed, as well as in the possibilities that exist to optimize this process (such as maximal speed under minimal material removal rate and minimal energy dissipation needed to fulfill its functionality). Traditional or equilibrium thermodynamics revolves around only the first and the last stages of spontaneously occurring processes. These are the conditions that produce a uniform distribution of energy, temperature, pressure, stresses, concentration, electric potential, *etc.* Generally, among the many fundamental laws of nature only the second law of thermodynamics points in the direction of time; the "arrow of time," in other words implies a direction and the possibility of evolution. That is why irreversible thermodynamics, as well as the theory of self-organization, is based on the second law of thermodynamics. Under non-equilibrium conditions, some transformations could take place during irreversible processes that cannot occur with equilibrium conditions, *i.e.*, at any temperatures, pressures, compositions, *etc.* For example, there are the processes that develop at entropy production decrease or at energy growth. The main concept of irreversible thermodynamics is that of entropy production. Entropy production characterizes the rate of entropy change when the irreversible process takes place in the system. The entropy production change characterizes the evolution of the process. Under stable conditions, the entropy production is simply related to the entropy flow, *i.e.*, to the entropy of the system change due to interaction with the environment by matter and energy. Materials removal or deformation could be characterized by exchange of matter and energy with the environment.

Materials removal rate develops under severe friction conditions (such as when cutting speed increases during machining) at the small scale. However, beginning with a definite value of the varying parameter (such as cutting speed), the materials removal rate could critically decrease. Under more severe working conditions the materials removal rate could drop down even further or at least differ slightly. A connection could occur between the technique characteristics, such as the materials removal rate decrease *vs.* cutting speed. Meanwhile, the structures originate within the system due to intensive matter and energy flow, *i.e.*, due to intensive irreversible processes. These structures are consistent with the energy flow and cannot exist without it. The structures that form under strongly non-equilibrium conditions are called dissipative structures and the process of dissipative structure formation is called self-organization. Dissipative structures differ from the equilibrium ones in the fact that they are the processes. A classic example of an equilibrium structure is a crystal. However, the thermal vibrations of atoms in the crystal are chaotic because these vibrations are not related or synchronized. If, under specific conditions, there is some interrelation in the vibration of atoms, then dissipative structures could form. Once the dissipative structures have been formed, the system (of the crystal for example) is able to transfer more energy practically without any damage, as compared to the conditions when these structures are not present. This could be explained by the following: the energy that has been spent on the system damage prior to the formation of the dissipative structures is now spent on their generation. That is why materials removal rate critically drops in MNT, owing to the initiation of the self-organization process. The self-organization during surface generation process in the MNT is usually associated with the process of the energy-rich surface structure formation. Entropy decrease and energy growth processes take place within these surface structures. If the phase and structure transformation takes place in a solid, then because the flows of matter and energy as well origins of these processes disappear, the dissipative structures disappear as well. However, the metastable structures still remain, and because the friction has been stopped, they are the secondary structures. The process of self-organization can start because the system has lost its stability.

To understand the features of the self-organizing phenomenon during cutting, one should understand the processes occurring at the surface of the "cutting tool-workpiece" system over a range of cutting speeds. Studies of the cutting of structural steel (~1040) have shown that tool life can vary dramatically with speed. At low speeds of cutting (up to 50 m/min), lying in the domain of cutting with HSS tools, intensive build-up formation takes place. The transformation of the machined material at the tool surface is frequently observed during metalworking and is common in structural steels. At the same time, oxygen (from the air) penetrates into the cutting zone. Fe-containing chip fragments will react with oxygen and carbon at the tool surface and form a boundary layer of both iron carbides and oxides. This leads to a built-up edge. The formation of a built-up layer is considered to be the result of self-organization of the tool-workpiece system and dissipative structure formation at low cutting speeds. The built-up layer is a dissipative structure or composite "third-body," which consists of heavily deformed and refined machining material, as well as oxides, nitrides, and other compounds generated during cutting. The built-up layer is similar in many ways to a composite material. The "ceramic-like" built-up layer offers significant protection to the tool surface. However, the stability of a built-up layer as a dissipative structure is very low, especially when cemented carbide tools are used under attrition wear conditions. For example, the adhesive interaction of tungsten carbide grains with a machined part can lead to microcrack formation. These cracks are generated at the interface of the phases because of the cyclical stress action at the points of adhesion with the workpiece, and this lead to separation of carbide grains from the tool surface. The principal mechanism of wear (and hence the dissipation of energy) of cemented carbide tool surface layers is the formation of microcracks and breaking off of the carbide grains. The failure of the built-up layer results in significant tool surface damage, as well as cutting edge breakage and chipping. The formation of a built-up layer demonstrates that the formation of a dissipative structure with low stability can also lead to increased wear if the friction surface does not ensure stable wear behavior of the dissipative structure formed. This is typical for cemented carbide cutting tools working under attrition wear conditions.

At cutting speeds higher than 50 m/min (used without a coolant), the process of seizure intensifies. On the surface of the tool face close to the cutting edge, a zone of plastic contact with a high friction coefficient will be formed. It results in the formation of a thin layer of heavily worked material at the tool-chip interface. The wear rate of the carbide tool is reduced with an increase in the cutting speed of up to 50 to 80 m/min. At optimum cutting speeds for carbide tools (50 to 80 m/min), the contact processes at the tool surface become nearly constant, whereas the formation of the built-up layer decreases and a flow zone forms. The formation of the flow zone with an increase in the cutting speed is the outcome of self-organization of the tool-workpiece friction pair, leading to a stabilization of friction. In this situation, the tungsten carbide grains undergo considerable fragmentation. This reduces the wear rate because of a decrease in the volume of the spalled fragments. With a further increase in cutting speeds (more than 100 to 150 m/min), the wear rate again increases owing to an intensification of the diffusion processes and the separation of carbide grains from the tool. These high cutting speeds lie in the domain of application of cermets and ceramic cutting tools, as well as surface-engineered cutting tools.

One of the principal features associated with cutting is a rapid increase in the dislocation density near the surface with the deformation being localized in a thin layer. Under extreme cutting conditions, this layer can undergo dynamic recrystallization, leading to a very fine-grained structure. This structure appears to be very stable during wear because it has the ability to both effectively accumulate and dissipate energy. The increase in dislocation density in this local volume is accompanied by activation of the surface layers of the tool, leading to further interactions with the environment. This can result in the formation of passivating surface structures (*i.e.*, solid solutions of oxygen in the metal or tribo-oxides), which can control any damaging adhesive interactions at the cutting tool-workpiece interface. Fragmentation of the carbides (or metal matrix) followed by subsequent formation of energy-absorbing, oxygen-containing surface film (at optimal cutting speeds) can also be considered to be an example of a cutting tool-workpiece self-organizing phenomenon, which occurs during the cutting-in stage.

The oxygen-containing films that are formed on the surface during machining have a high energy accumulation capacity, which can improve the tool life. Under low-cutting-speed conditions, oxygen-containing films such as Fe-O, Mo-O, and W-O are formed for the cutting tools made of HSS, whereas W-O, Co-O, and Ti-O films are formed for cutting tools made of cemented carbides. Under optimal cutting conditions for cemented carbide tools, high-oxygen content compounds such as WO_3, TiO_2, and Ti_2O_3 can occur. The elevated values of the energy accumulation capacity of Ti_2O_3 oxides is possible one of the reasons for the improved tool life of TiC-containing carbides compared to WC ones.

As most of the cutting works are completed by the local cutting edge in MNT which makes most of the phenomenon belong to micro-tribology area (wear). In the common sense, MNT can be assumed as an integrate process of friction, wear and lubrication (tribosystem). In the case of MNT, surface secondary structures perform protective functions, limiting the interactions (occurring during friction) into the depth of cutting tool (workpiece) and reducing the intensity of such interaction. In fact we can assume that the formation of secondary structures is a system's response to an external impact, *i.e.*, friction. This reaction is aimed at reducing the external impact by decreasing materials removal rate. If the system cannot form stable secondary structures, it tries to stop the friction by means of strong seizure or jamming. Furthermore, surface films (or secondary structures) are necessary for energy dispersion during its transformation from a friction zone into a frictional body. To obtain minimal materials removal rate (wear rate), a dissipation of energy should occur under the lowest rate of entropy growth.

The independence of the nature of dissipative structures and surface films (secondary structures) from the entry conditions is related to the fact that a process of self-organizing begins after the system loses the stability of its previous state. The system can pass through this instability several times because of evolution. An increasing external effect, *e.g.*, the increase in sliding speed or frictional force, causes a systematic deviation of a given system from the equilibrium state, and therefore dissipative structures are able to form within the system. In the first instance, it loses its stable equilibrium state with the occurrence of a relative nonzero sliding speed. This can be attributed to the fact that, in the equilibrium state, the macroscopic particles of the system cannot move against each other. Further losses in stability take place owing to the evolution of secondary structures on the friction surfaces.

In real systems some sources of energy dissipation can be available along with friction, such as various oscillation processes. In addition to friction, some physical-chemical interaction always takes place between frictional pair (workpiece-tool), the environment, and the lubricant. With significant deviations from an equilibrium state, the specific chemical kinetics will also be important in providing the stability of the given state of a system. In reality, physical and chemical interactions between frictional body with the environment or a counter-body results in formation of surface films. The properties of superficial secondary structures considerably differ from the properties of a frictional body. Heat conductivity will change during machining process because of the formation of surface secondary structures, *i.e.*, the change of heat conductivity will characterize physical and chemical processes occurring on the surface. Then, in the expressions signifying excessive entropy production, not only the coefficient of friction but also heat conductivity values will vary. As the friction is the only independent source of energy dissipation in the system, the excessive entropy production has the following forms:

As the load varies:

$$\frac{\partial(\delta^2 S)}{2\partial t} = (v^2/BqT^2)\left(p\frac{\partial k}{\partial p}+k\right)\left(\left(p\frac{\partial k}{\partial p}+k\right)-(kp/q)\frac{\partial q}{\partial p}\right)(\delta p)^2$$

$$= (v^2/BqT^2)\left(\left(p\frac{\partial k}{\partial p}+k\right)^2 - (kp/q)\left(p\frac{\partial k}{\partial p}\frac{\partial q}{\partial p}+k\left(\frac{\partial q}{\partial p}\right)^2\right)\right)(\delta p)^2$$

(5.1)

As the velocity varies:

$$\frac{\partial(\delta^2 S)}{2\partial t} = (p^2/BqT^2)\left(v\frac{\partial k}{\partial v}+k\right)\left(\left(v\frac{\partial k}{\partial v}+k\right)-(kv/q)\frac{\partial q}{\partial v}\right)(\delta v)^2$$

$$= (p^2/BqT^2)\left(\left(v\frac{\partial k}{\partial v}+k\right)^2 - (kv/q)\left(v\frac{\partial k}{\partial v}\frac{\partial q}{\partial p}+k\left(\frac{\partial q}{\partial v}\right)^2\right)\right)(\delta v)^2$$

(5.2)

Excessive entropy production in equation 5.1 and equation 5.2 can become negative because of a negative sign "-" before the second term in the brackets. Whereas excessive entropy production could become negative, the sum $\left(p\frac{\partial k}{\partial p}\frac{\partial q}{\partial p}+k\left(\frac{\partial q}{\partial p}\right)^2\right)$ or $\left(v\frac{\partial k}{\partial v}\frac{\partial q}{\partial p}+k\left(\frac{\partial q}{\partial v}\right)^2\right)$ should remain positive. It is defined by the sign of the products $\frac{\partial k}{\partial p}\frac{\partial q}{\partial p}$ and $\frac{\partial k}{\partial v}\frac{\partial q}{\partial p}$. The friction pair in MNT can lose its stability if the coefficient of friction and heat conductivity simultaneously increase or decrease with loading or speed. The increase of the coefficient of friction with loading or speed can result in a seizure. The increase of heat conductivity will be observed in this case. It can be called the degenerative case of self-organization when the system tries to stop the friction and wear process with the help of seizure (there is no friction within a zone of seizure). Reduction of the coefficient of friction with the loading or sliding speed is associated, as a rule, with the formation of surface secondary structures and corresponds to the process of self-organization.

Generally, it is possible to make a very important conclusion that the process of self-organization during machining is possible if one more or more independent process, except friction is affecting a body. The analysis of known self-organizing systems shows that dissipative structures form as a result of the interaction of two or more processes. For example, the formation of the Benard cells is the result of interaction of thermal conductivity and gravity. Turing structures are formed as a result of interaction between chemical reaction and diffusion. Martensite formation is the result of the interaction of thermal conductivity and phase transformation. On the other hand, thermo-diffusion and similar phenomena are not the result of self-organization, because they are caused by a single process. They are triggered by this specific process and develop gradually without sudden change. Therefore, the system does not lose its thermodynamic stability. Under varying conditions, self-organization could accelerate or slow different processes. To be initiated, self-organization needs two or more independent external impacts on the system. Moreover, this conclusion could be expanded for other physical processes. The oxidation rate could be controlled by self-organization if a second non-equilibrium process exists in addition to the isothermal oxidation. This is a relaxation process within strongly non-equilibrium nanostructures.

Self-organization and formation of dissipative structures take place in a nonlinear area. Therefore, Curie's principle is not operative in this area. The processes described by tensors of ranks of different parity can cooperate inside dissipative structures. For example, the speed of chemical processes is a scalar, *i.e.*, it is a tensor of a zero rank; the thermal flow is a vector, *i.e.*, it is a tensor of the first rank. These processes do not interact with each other on a linear area, but are able to interact within dissipative structures. In a stationary state of the dissipative structures, minimal entropy production keeps the value within a wide range of changes of external conditions. Strictly speaking, Prigogine's theorem does not operate under the conditions of the existence of the dissipative structures. However, according to the Klimontovich theorem, entropy production after self-organization will be lower than at previous conditions. Therefore, it is possible to assume that, for the new stationary condition Prigogine's theorem can be applied. In this case, it is possible to neglect the flows of unfixed thermodynamic forces. Thermodynamic forces that cause a flow of matter are not fixed. Therefore, the terms in the equation for entropy production connected to flows of matter, *i.e.*, the wear process, can be neglected. Let us assume that entropy production does not vary within the certain range of loading and speed of sliding change and it will follow from:

$$k^2 = CBqT^2 / (pv)^2 \tag{5.3}$$

here C is the constant value of entropy production. As described in equation 5.3, in a stationary state of dissipative structures, at minimal materials removal rate, relations could also appear between a coefficient of friction and loading and sliding speed at the tool-chip interface. In this case, the coefficient of friction can decrease with values of loading and speed of sliding.

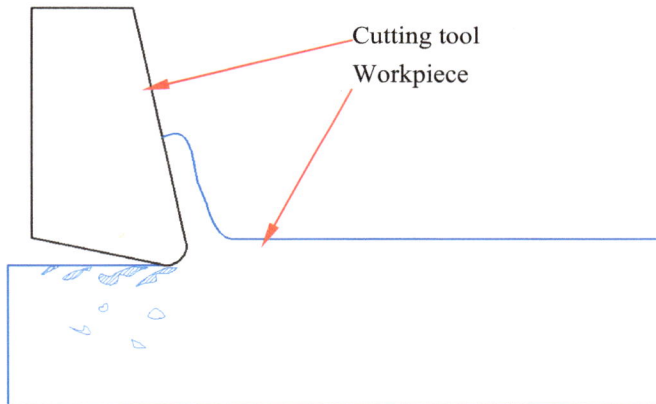

Figure 1: Surface self-organization in MNT.

Mechanisms of these relations could be different, and they are usually connected to nonequilibrium processes. These processes can be non-equilibrium chemical reactions in which the material of a frictional body cannot participate as a reagent. It can be partition of phase components, such as allocation of a soft phase in a surface layer. In these materials, the soft phase had a uniform distribution in the structure of the alloy. Owing to the occurrence of the aforementioned relations the soft phase concentrates within a surface layer. After that the distribution ceases to be equal and so an ordering of structure occurs. It should be accompanied by a decrease in entropy. We can assume that, at the very moment when the dependence of the coefficient of friction at the tool-workpiece interface on other friction parameters occurs, it can be used as a criterion of self-organizing process initiation and formation of dissipative structures.

Example of structure self-organization in the micro & nanomachining is shown in Fig. (**1**). This is some kind of two-phase alloy with a hard matrix and evenly distributed soft inclusions. The self-organization is developed by the accumulation of the soft phase on the machined surface. Taking into account the independence of dissipative structures from entry conditions, *i.e.*, their "constancy" under the given conditions of friction, it is obviously possible to perform the development of different anti-wear cutting tool materials based on the investigations of the characteristics of surface films (secondary structures). If the materials of the frictional bodies have a nonequilibrium structure, the process of relaxation is enhanced. In this case, the relaxation process interacts with the process of friction. Interaction of these processes could lead to a significantly higher probability of initiation of the self-organization process of a frictional body with a nonequilibrium structure than that of an equilibrium one.

The self-organization process during machining can possibly be used in the proper selection of tooling materials, as well as for the development of surface-engineering techniques. More attention will be given to specific tooling materials with enhanced adaptation ability. These adaptive materials constitute a promising generation of novel materials. The fabrication of these materials can be accomplished through surface engineering and powder metallurgy treatments.

2. DYNAMIC EQUILIBRIUM MECHANISM BETWEEN DIFFERENT SCALES

Research of MNT is rapidly advancing owing to the combined efforts of science and engineering through modeling and simulation together with experimentation. Computational modeling, in particular atomistic and molecular simulation, is becoming increasingly important in the development of MNT. At nanoscale, the effects of single atoms, individual molecule, or nano-structural features may dominate the material behavior. Thus novel modeling and simulation approaches are a vital component in enabling the manufacturing process. Atomistic models, by providing a material description that starts at a fundamental scale, are thus expected to be important not only for scientists but also for engineers. Besides molecular dynamics, another emphasis is on developing a sensitivity for the significance of mechanics in different areas, and how atomistic and continuum viewpoints can be integrated to build a new platform of control in the analysis and description of the behavior of complex materials under extreme conditions. Atomistic modeling of small-scale dynamics phenomena with large-scale simulations combined with theoretical methods such as continuum mechanics are the meaningful method on MNT investigation.

Material deformation (removal) in the MNT is a phenomenon that cannot be understood at a single scale alone. It requires the consideration of multiple scales to capture the progression of the elementary physical mechanisms. How can one capture multiple scales? One possibility is to simply simulate all process in a system. Another possibility is to use a combination of methods with different accuracy or resolution. This idea is motivated by the fact that in many problems, a high resolution and a high accuracy is only required in regions that are small compared with the overall specimen size. This is for instance the case in the example of a crack-like defect, where only atoms close to the crack tip experience large stresses, whereas atoms further away undergo only small deformation. The reason why there is no single, perfect and all-comprising model for calculating material properties on all scales relevant for MNT, is, that nature exhibits complex structural hierarchies which occur in engineering systems as well as in biological systems (self-organization of matter), investigated in the life sciences. Remarkably, on the angstrom scale there are only atoms but then on larger length scales these basic constituents build complex hierarchical structures in biology and engineering, which are treated with different theories that have a certain range of applicability. Materials of industrial importance such as glasses, ceramics, metals or (bio) polymers, today are increasingly regarded as hierarchical systems and there has been a focus of research on the investigation of the different structures of components, of classes of materials on various structural hierarchies, and their combination within "process chains". Hence, the typical structural or architectural features of materials on different scales have to be taken into account. Usually one of two possible strategies is pursued. In a bottom-up approach the many degrees of freedom on smaller scales are averaged out to obtain input parameters relevant on larger scales. In contrast, a top-down approach tries to establish the structure on smaller scales starting from a coarse-grained level of resolution.

The increased computational power and programming capabilities in recent years have given impetus to the multi-scale modeling, which implements the largely intuitive notion that physical phenomena occurring at a lower length or size scale determine the observed response at a higher scale. A logical outcome of this thought is an organization of differentiated scales-from the lowest, such as nanometer scale, to the highest scale typical of the part or structure in mind-giving a hierarchy of scales. Working up the scales produces a hierarchical multi-scale modeling, in which the essential challenge consists of "bridging" the scales. The simulation techniques, such as molecular dynamics simulation (MDS), succeed mostly in revealing phenomena from one scale to the next; but proceeding to three or more scales often necessitates unrealistic computing power even with the most versatile facilities available. In addition, the limitation of independent physical validation of the simulated results questions the wisdom of total reliance on the multi-scale hierarchical modeling strategy. When it comes to subcritical (pre-failure) damage in materials, the multi-scale modeling concept needs closer examination, firstly, because the length scales of constituents and heterogeneities are fixed while those of damage evolve progressively, and secondly, because the mechanisms of damage tend to segregate in modes with individual characteristic scales. All this is the subject of this section, which will first describe and clarify the damage mechanisms in common types of materials followed by the induced response observed at the macro-scale. The hierarchical modeling approach will be discussed against this knowledge; and a different approach, named synergistic multi-scale modeling, will be advocated.

Generally speaking, the materials removal (plastic deformation) occurs due to a variety of dissipative mechanisms which cause permanent changes in the internal microstructure of the material and decrease the energy storing capacity of the material. The most basic scale at which these mechanisms occur depends upon the size of inhomogeneities in the microstructure of the material. As an example, materials may show dissipative mechanisms at the nanometer scale. In reality, however, identifying this scale is limited by the ability to observe as well as to model and analyze the mechanisms at the observed scale. The so-called microscale is a reference to the scale at which entities or features within a material are observable by a certain type of microscope. Thus, for example, the microscale can be a few micrometers, if an electron microscope is used to observe entities, such as cracks or crystalline slip within grains or at grain boundaries. The scale reduces by an order of magnitude if one focuses on dislocations observed by a transmission electron microscope. Today, the use of nanoscale elements (particles, fibers, tubes, *etc.*) has moved the basic scale further down to the atomic scale. At this scale, the basic notions of continuum mechanics fail; and it is necessary to develop modeling tools that can bridge the discrete-level descriptions (quantum mechanics) to continuum type (smeared-out) descriptions.

In an engineering approach, the purpose at hand should guide the choice of the basic scale. Thus, if the overall (effective) characteristics of inelastic response are of interest, it would suffice to incorporate the energy dissipating mechanisms in a model, directly or indirectly, in an appropriate average sense; while if, for instance, the aim is a particular material failure characteristic, the analysis may need to be conducted at the local physical scale of the relevant details of the mechanisms. On the other hand, if the purpose is to design a material, *i.e.*, to engineer its response or to provide it with certain functionalities, then it would be necessary to address scales where the material (micro) structure can be modified, manipulated, or intruded. In MNT, the scales of inhomogeneities (reinforcements, additives, second phases, *etc.*) embedded in the baseline material (matrix) determine the characteristic scales of operation of the mechanisms of energy dissipation. Although energy dissipation may also be occurring at other (smaller) scales, *e.g.*, the scale of the matrix material's microstructure, and the dissipative mechanisms associated with the inhomogeneities have usually an overriding influence on the materials deformation behavior. The complexity introduced by inhomogeneities in materials removal (deformation) is in the form of multiple scales of dissipative mechanisms depending on the geometrical features of the inhomogeneities.

Multiscale method enable us to investigate the physical phenomenon based upon complex interaction of different scales [1-2]. In many this type systems, some local area should be simulated using atomistic modeling method with high precision which should take long period of computation while some general precision method such as continuum method is enough about study some region. For example, in the MNT the real cutting process is happened at local area around tool tip which means a high precision method is necessary while some region far away from the tool tip can be investigated using finite element method. There are two main advantages of using this methodology, on one hand it can shorten the simulation time and on the other hand it can ensure the seamless transmitting of thermodynamics and mechanics information and thus result in precise description of real physical system in a short while. General machining technique can be studied within some definite length scale while some different properties exist in some specific application such as phase transformation and MNT. The simulation model of those applications is too large for atomic length scale theory but is too small for macroscopic length scale method.

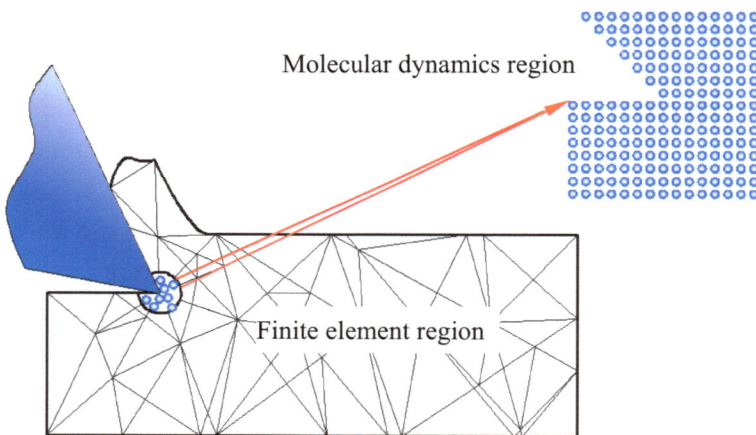

Figure 2: Multiscale model for MNT.

Fig. (**2**) gives the typical multiscale model for MNT. The atomic bond around tool tip is bound to fracture for the materials removal which is characterized by the obvious variation of electronic density at angstrom length scale. The fracture of atomic bond leads to the severe lattice distortion and the corresponding amorphous structure. There is only minor deformation about the crystal lattice in the local region away from the tool tip. Crystal deformation of substrate materials is observed within definite distance (about nanometer length scale) away from tool tip. There is almost no crystal lattice deformation in the region farther away (about micrometer length scale) from tool tip which is controlled by continuum mechanics. Presently, coupled interaction between the last two length scales is widely used in MNT investigation as it ensures enough precision which is also the center point of this section. In reality, the cutting process resulted by materials fracture can be simulated by corresponding computation model in all these three length scales. The quantum mechanics model based upon electronic cloud can be used to study the physical process lies within angstrom length scale away from tool tip. The crystal deformation at nanometer scale can be modeled using empirical potential function while the materials deformation at micrometer scale can be modeled by elastic-plastic deformation and finite element method.

The physical phenomena at different length scales seamlessly coupled together and thus form a systematic ensemble. The fracture process at the local area around tool tip has intense relationship with the deformation mode and the crack transmission in adjacent area and the chemical bond fracture around the tool tip will influence the deformation of the adjacent area *vice versa*. The transmission of the microcrack at macroscopic scale is induced by the stress relaxation while the strain energy is released by some local dynamic process such as cutting heat, plastic deformation and the elastic wave emission. The time duration from tool tip to adjacent area needed for elastic wave transmission is about nanosecond level as the elastic wave is transmitted with sound speed. All of the above processes justify that the reasonable couple method for dynamic equilibrium is necessary for a convinced simulation of MNT.

The coupling effect is not strong in some specific applications such as the investigation of thermal conductivity of silicon can be analyzed using hierarchical modeling method. The computational parameters can be transferred among different scales through appropriate setting by which the computation results of the small scale can be exerted as input loading conditions of the larger model and the computation results of the larger scale can be assumed as boundary conditions of the small scale. Here the scale refers in particular to length, time or energy. The hierarchical coupling is mainly used to applications where definite computation precision can be acquired at single scale. The larger scale is smoothly integrated by numerous small scales which the larger scale can be assumed as continuum and quasi-static. Recently, the research on concurrent coupling method gradually increased with the development of deep coupling system such as micro-electro-mechanical system (MEMS), micro/nano mechanism and lab on chip (LOC) and so on. The common feature of the above system is the inherent multiscale property which means the property of every scale will depend upon the behavior of other scales. The scientific model is necessary for reasonable describe the transition area to ensure the continuity among different regions. The concurrent coupling method is still a new region of scientific research while few widely accepted model is provided. There are not strict rules about how bridging different scales but some common view does exist. It is necessary to set the reasonable boundary conditions to ensure the continuity at the interface as the physical structure of different scales being seamlessly connected to each other. In general, smooth transition ought to be guaranteed and comply with dynamic property and conservation law of different scales. The thickness of the transition area should as small as possible which can reduce the corresponding parameter.

As for MNT, nanoscale mechanism dominant the materials removal region around tool tip while the microscale discipline control materials elastic-plastic deformation and fracture adjacent to this region. It is necessary for the coupling model of these scales to ensure the continuity of materials static elastic response and also the dynamic elastic response (phonon) as the successively emitted of elastic wave from tool tip. Furthermore, the continuity of thermal-dynamic property ought to be ensured because of continuously generating and dissipating of cutting heat.

In the bridging-scale method originated by Wanger and Liu [3], the atomic displacements are decomposed into coarse-scale and fine-scale displacements. The decomposition of the matrix of atomistic displacements is:

$$\mathbf{u}_I^A = \mathbf{N}_{IJ}\mathbf{d}_J^C + \tilde{\mathbf{u}}_I^A$$

$$\mathbf{u}_I^A = \mathbf{N}\mathbf{d}^C + \tilde{\mathbf{u}}^A \qquad (5.4)$$

$$\mathbf{N}\mathbf{d}^C = \overline{\mathbf{u}}$$

Here $\overline{\mathbf{u}}$ is the coarse-scale field, which is usually a finite element continuum field; $\mathbf{d}^C$ is the nodal displacement of a coarse scale mesh. The key concept is that the coarse-scale displacement is obtained by a least square projection weighted by the atomic masses which the projection is

$$\mathbf{d}^C = \mathbf{M}^{-1}\mathbf{N}^T\mathbf{M}^A\mathbf{u}^A \tag{5.5}$$

here $\mathbf{M}$ is the matrix of the coarse - scale masses given by

$$\mathbf{M} = \mathbf{N}^T\mathbf{M}^A\mathbf{N} \tag{5.6}$$

which is the mass that would be obtained by a master-slave relationship. It can be seen from equation (5.4) that the nodal displacement of the finite element mesh can then be obtained at any time from atomistic displacements. Combining (5.4) and (5.5) gives

$$\mathbf{u}_I^A = \mathbf{N}\mathbf{M}^{-1}\mathbf{N}^T\mathbf{M}^A\mathbf{u} + \tilde{\mathbf{u}}^A = \mathbf{P}\mathbf{u} + \tilde{\mathbf{u}}^A \tag{5.7}$$

$$\mathbf{P} = \mathbf{N}\mathbf{M}^{-1}\mathbf{N}^T\mathbf{M}^A \tag{5.8}$$

It follows from the above that

$$\tilde{\mathbf{u}}^A = (\mathbf{I} - \mathbf{P})\mathbf{u}^A \tag{5.9}$$

In the implementation of Wanger and Liu [3], this projection is apparently only used at the interface. By using this concept, they are able to obtain an equation for the fine scale at the interface without any Lagrange multipliers. The approach has been improved and then streamlined by updating the interface atoms by Green's functions for lattice. Furthermore, this method also entails the computation of Fourier transforms at the interface.

3. THERMODYNAMICS OF MNT

As the depth of cut and materials removal volume is very small in MNT, most of the cutting works are finished by tool tip which result in the different interaction at the tool-chip interface and materials plastic deformation. Different control mechanisms in MNT lead to different thermal dynamics property. Many of the materials removal processes described above involve plastic deformation of material in small localized regions where asperities of the opposing surfaces or hard particles make contact. It turns out that the onset of plasticity for such small regions can be dramatically different than the plastic yield stress determined by macroscopic measurements. Macroscale plastic deformation within crystalline materials occurs mainly through the motion of flaws and defects in the crystal structure, called dislocations. When the size of the region undergoing plastic flow becomes much smaller than the typical distance between dislocations, however, the contribution of dislocations to the yield stress becomes insignificant; instead, the yield stress is governed by the force needed to slide one plane of atoms over another. For a perfect crystal with no dislocations, the theoretical shear stress needed to slide a plane of atoms over another in MNT is far greater than that of common machining which result in more intense atomic thermal motion.

Mechanical properties of materials are scale dependent based on the strain gradient plasticity and the effect of dislocation-assisted sliding. When material is removed by machining there is substantial increase in the specific energy required with decrease in chip size. It is generally believed this is due to the fact that all metals contain defects (grain boundaries, missing and impurity atoms, *etc.*), and when the size of the material removed decreases, the probability of encountering a stress-reducing defect decreases. Since the shear stress and strain in machining is unusually high, discontinuous microcracks usually form on the metal-cutting shear plane and the heat generation at shear plane is different from common machining process.

Different heat generation process will lead to different effect upon materials surface integrity as follows:

Diffusion leading to grain growth and softening.

Phase transformations leading to re-hardening.

Thermal effects leading to expansion, contraction, possible cracking, and tensile residual stresses.

Chemical reactions leading to increased oxidation.

Heat transfer in devices and structures on the micro- and nanoscale is a broad field and is treated extensively by many researchers. The first Fourier's law of heat transfer describes the correlation between the steady heat flow and the driving temperature difference, here for one-dimensional heat conduction:

$$\dot{Q} = -\lambda A(r)\frac{dT}{dr} \qquad (5.10)$$

The integration over the coordinate r leads for constant heat conductivity λ to

$$T_1 - T_2 = \frac{\dot{Q}}{\lambda}\int_{r_1}^{r_2}\frac{1}{A(r)}\,dr \qquad (5.11)$$

The integrated area divided by the thermal conductivity is often called the thermal resistance in analogy to electrical resistance. The heat conductivity in MNT is influenced by the microstructure of the material and the working conditions. Grain boundaries and crystal lattices form additional resistances to the heat transfer. In regular crystals, the heat transfer coefficient is dependent on the crystal orientation, and the Fourier equation of heat transfer must be expanded to the tensor notation

$$\dot{q} = -\lambda\frac{dT}{dx} \quad \rightarrow \quad \dot{\vec{q}} = -\Lambda\,\text{grad}T \qquad (5.12)$$

with the heat conductivity tensor

$$\Lambda = \begin{pmatrix} \lambda_{11} & \lambda_{12} & \lambda_{113} \\ \lambda_{21} & \lambda_{22} & \lambda_{23} \\ \lambda_{31} & \lambda_{32} & \lambda_{33} \end{pmatrix} \qquad (5.13)$$

The solution of the three-dimensional heat conduction is often only possible with numerical methods. CFD and FEM programs allow the treatment of complex heat transfer problems. The time dependent second Fourier law is derived from a differential element with the balance of the heat capacity and the heat conduction

$$\rho A(r)c_p\frac{\partial T}{\partial t} = \frac{\partial}{\partial r}\left(\lambda A(r)\frac{\partial T}{\partial r}\right) \qquad (5.14)$$

Constant material properties, such as heat capacity c_p and heat conductivity λ, and mathematical simplification for different geometrical bodies ($n = 0$ for a plate, $n = 1$ for a cylinder, and $n = 2$ for a sphere) lead to

$$\frac{\partial T}{\partial t} = a\left(\frac{\partial^2 T}{\partial r^2} + \frac{n}{r}\frac{\partial T}{\partial r}\right) \tag{5.15}$$

with the temperature conductivity or heat diffusivity $a = \lambda / \rho\, c_p$. The transient temperature development in a semi-infinite body, is given in one-dimensional form by

$$\frac{\partial T}{\partial t} = a\frac{\partial^2 T}{\partial x^2} \quad t \geq 0 \; ; \; x \geq 0 \tag{5.16}$$

The introduction of the dimensionless temperature θ leads to

$$\frac{\partial \theta}{\partial t} = a\frac{\partial^2 \theta}{\partial x^2} \text{ with } \theta = \frac{T - T_0}{T_W - T_0} \tag{5.17}$$

describing the temperature development in a solid body with defined wall temperature T_W. The solution of Eq. 5.17 is dependent on the boundary condition at the wall. With constant wall temperature $T_W = $ const., the dimensionless temperature is determined by the error function.

$$\frac{T - T_0}{T_W - T_0} = \text{erf}(x^*) \text{ with } x^* = \frac{x}{2\sqrt{at}} \tag{5.18}$$

and the Gaussian error function

$$erf(x) = \int_0^x e^{-\varsigma^2} d\varsigma \tag{5.19}$$

The wall heat flux can thus be calculated from Eq. 5.18. For a constant heat transfer coefficient a at the wall, the solution of Eq. 5.17 is derived with the help of two dimensionless numbers, the Fourier number

$$\text{Fo} = \frac{at}{x^2} \tag{5.20}$$

and the biot number

$$\text{Bi} = \frac{\alpha x}{\lambda_s} \tag{5.21}$$

The Bi number is similar to the Nußelt number Nu from convective heat transfer, except for the heat conductivity of the solid body. The Fourier number is also defined for mass transfer problems Fo' = $D_m t / x^2$ with the molecular diffusion coefficient D_m. The temperature development during cooling of the body is given by

$$\theta_C = \frac{T - T_\infty}{T_0 - T_\infty} = \text{erf}(x^*) - e^{\text{FoBi}^2 + \text{Bi}} \ \text{erfc}\left(\sqrt{\text{Fo}}\,\text{Bi} + x^*\right) \tag{5.22}$$

and for heating

$$\theta_H = 1 - \theta_C \tag{5.23}$$

The characteristic time for heating or cooling is drawn from the combination of the Fo and Bi numbers and depends not on the length (semi-infinite body). A miniaturization will not influence the temperature development and the heat flux for a semi-infinite body.

$$t_C = at\left(\frac{\alpha}{\lambda}\right)^2 = \text{FoBi}^2 \tag{5.24}$$

The unsteady heat transfer in MNT at local area is rapid and can often be approximated with the fast asymptotic behavior of a steady state solution. For bodies (high Fo and low Bi number), the temperature distribution can be approximated by asymptotic solutions. The temperature inside a small area is only a function of time

$$\theta = \frac{T - T_\infty}{T_0 - T_\infty} = \exp\left(-\frac{\alpha A t}{m c_p}\right) \tag{5.25}$$

while the characteristic time is

$$t_C = \frac{mc_p}{\alpha A}$$ (5.26)

With the temperature diffusivity of the solid $a = \lambda / \rho c_p$ and the characteristic length $l \propto A^{0.5}$, this equation can be rewritten

$$t_C = \frac{l^2}{aBi}$$ (5.27)

3.1. Heat Conduction in Fixed-Abrasive Machining

As for fixed-abrasive machining, it would appear sensible to present temperature results for various process variables such as wheel speed and work speed. However, no sensible interpretation of such results can be obtained without reference to the accompanying values of specific energy. This should now be clear from the above examples. To illustrate this point, consider the case where grinding wheel speed is increased. If the effect of the increased wheel speed is to increase specific energy, the workpiece temperature will be increased. If the effect of the increased wheel speed is to reduce specific energy, the workpiece temperature will be reduced. For a more complete understanding of the effect of any process variable, it is valuable to examine more closely the heat transfer in the region of the abrasive contact. Some of the heat conduction in the workpiece is investigated in the following section. The heat flux, q_w, conducted into the workpiece is only a proportion of the total heat flux, q_t. Defining this proportion as R_w, the heat flux which enters the workpiece is

$$q_w = q_t R_w$$ (5.28)

Since q_t varies as a function of position in the contact zone, as in Eq. (5.28), we can write more generally

$$q_w(x) = q_t(x)R_w$$ (5.29)

Typically, the workpiece partition ratio, R_w, varies according to the type of abrasive, the workpiece material, the specific energy, the grinding fluid, and the contact length. Grinding with an alumina wheel in a dry shallow-cut operation, the value of R_w may be as high as 90%, whereas in well-lubricated creep-feed grinding the partition ratio may be less than 5%. The band of heat in a practical grinding contact is usually considered as a series of moving line source elements. The solution for the moving line source can be obtained using Bessel functions which are available from tables or as functions in mathematical software. The line source temperatures must be summed over the length of the grinding contact. This summation process is expressed by the integrals in Equation (5.29) and (5.30) assuming a semi-infinite workpiece. The band heat source is usually considered to lie in the flat plane where $z = 0$. For this case, the solution for a band heat source moving in the x-direction with flux, q, varying in strength with position, a, within the contact length is given by

$$T = \int_{-\frac{l_c}{2}}^{\frac{l_c}{2}} \frac{q}{\pi k} e^{\frac{v(x-a)}{2\alpha}} K_0 \left\{ \frac{v\left[(x-a)^2 + z^2\right]^{\frac{1}{2}}}{2\alpha} \right\} da \tag{5.30}$$

where $K_0[u]$ is the modified Bessel function of the second kind of order zero for an argument of value u. The integration variable, a, defines various positions of the line source elements within the heat band.

$$-\frac{l_c}{2} \leq a \leq \frac{l_c}{2} \tag{5.31}$$

For deep cuts, as experienced in creep-feed precision grinding, the moving heat should be expressed in terms of the angle, φ, between the line of motion and the plane of the band source. For the inclined band source, the equation to be solved is

$$T = \int_{-\frac{l_c}{2}}^{\frac{l_c}{2}} \frac{q}{\pi k} e^{\frac{v(x-a)}{2\alpha}} K_0 \left\{ \frac{v\left[(x-a\cos\varphi)^2 + (z-a\sin\varphi)^2\right]^{\frac{1}{2}}}{2\alpha} \right\} da \tag{5.32}$$

The thermal properties of the workpiece material are represented by k, ρ, c, and α, where k is the thermal conductivity, ρ is the density, c is the specific heat capacity, and

$$\alpha = \frac{k}{\rho c} \tag{5.33}$$

is the diffusivity. For special cases, a simpler approach can be employed as shown in later sections. However, Equation (5.31) and (5.32) give a more rigorous analysis of workpiece temperature.

4. ENERGY DISSIPATIVE MECHANISM OF MNT

4.1. Macroscopic View

Energy dissipation in MNT is an important issue although we know little about the physical essence of this process. In reality, energy dissipation in MNT is mainly realized through process of elastic deformation or materials fracture or tribology behavior. The process by which the mechanical work is not transmitted through a mechanical system is, of course essentially irreversible. Since they are the ultimate origin of friction, it is necessary to carefully analyze the process in order to gain a physical picture of the effects that take part in the dissipation of mechanical energy in MNT. Owing to the great complex of this irreversible process it is only possible so far to outline a qualitative picture of the main feature of the process. The whole course of the "loss" process of mechanical energy in MNT can be divided into three phases:

❖ Storage

  generation of point defects and dislocations

  strain energy storage

❖ Emission

  phonons

  electrons

❖ Transformation to thermal plane

  generation of heat and entropy

In the following, the main process of the dissipation will be introduced:

4.1.1. Storage of Energy

In the middle of 1950's it became possible to study the molecular array of the crystalline bodies and the dislocations and imperfections in this array by using of Moire technique in conjunction with an electron microscope. Since then, the plastic deformation and shearing of solids is governed by imperfections and dislocations and friction processes involve plastic deformation process. In 1965 Kostetsti and Nazarenko [4] tried to explain the Coulomb-Amontons law of sliding friction (tool-chip and tool-machined surface interface) through a relation between normal forces, friction forces and the dislocation structure of the solids. From the theory of dislocations it follows that the work per unit length to create a dislocation in a isotropic undisturbed medium is given by

$$\Delta E_e = \frac{Gb^2}{4\pi(1-v)} \ln\left(\frac{r_1}{r_0}\right) \text{ for edge dislocation} \tag{5.34}$$

$$\Delta E_s = \frac{Gb^2}{4\pi} \ln\left(\frac{r_1}{r_0}\right) \text{ for screw dislocation} \tag{5.35}$$

here G is the shear modulus, b is the Burgers vector, r_1 is the core radius, r_0 is the radius of influence, v is the poisson ratio. Take an example of copper, it was found that the stored elastic energy in a condition of incipient plasticity can only account for 1% of the energy expended in friction. A similar result is also found in conditions of more severe plastic deformation conditions of sliding. This may be due to the fact that in the plastic deformation mode the dislocations generated would not have remained in the surface after the passage of sliding counter face and the energy would appear as heat. Some part of the frictional energy is stored in the sliding partner *via* dislocation in the MNT. As the stored energy is only a small part of of the whole energy, other dissipation mechanism must also be studied.

4.1.2. Emission of Energy

In the microscopic model of MNT it has been explained that during sliding, micro-contact are formed and destroyed. These processes are statistically distributed in time and location within the geometric area of contact interface between tool and workpiece. It is likely that in the process of junction formation some of the asperities or some part of them deformed elastically. If then the junctions are destroyed and the adhesion bonds are broken, the elastic deformed parts of the asperities snap apart thus generating vibrations and other lattices stimulations which lead to the emission of energy in different forms. In solid state physics, different processes of transformation of energy into apparently other forms are known (such as photoelectric effect and thermo-ionic effect). In general, the main energy emission effect is tribology induced.

4.1.2.1. Emission of Phonons

The generation of acoustic waves and the emission of sound is a common feature of most MNT process. Since noise is considered as a "acoustic pollutant" the reduction of noise emission is an important task in today's manufacturing technology. Physically, the generation of sound in MNT is connected with process of elastic deformation and releases of asperities. A theory of natural normal micro-vibrations in sliding contact was put forward by Tolstoi [5]. He showed that the self-excited micro-vibrations during sliding are invariably accompanied by simultaneous upward jumps of asperities. The frequency of the natural micro-vibrations is determined by the contact stiffness and the mass of sliding surface. These micro-vibrations vanish only when the normal vibrations are damped by external means or when the velocity is lower than a certain limit. The value of the critical velocity can be roughly estimated as follows:

$$V_c = \frac{f\sigma_y l_a^2}{2\eta y} \tag{5.36}$$

here η is the creep viscosity of the solids, σ_y is the yield strength, f is the friction coefficient, y and l_a are the height and spacing of the surface asperities. As for the traditional steel the V_c is about $5\times10^{-13}\sim5\times10^{-12}$ m/s. The very low value indicates that in almost any sliding situation a certain amount of the mechanical energy is dissipated through the generation of vibration and the following emission of acoustic waves. The absolute amount of energy emitted   is only a very small part of the whole sliding friction energy.

4.1.2.2. Emission of Photons

The emission of photons occurs if certain solids are sliding against each other. Under this circumstance, the mechanical work which separates adhesion bonds activates photons which are emitted as visible light. A complete theoretical explanation of this effect is still needed. Some observations support the assumption that in the surface cracks of the crystals, electrostatic double-layers exist which cause a kind of spark discharging if the surface cracks are cleaved during the manufacturing process. Obviously some part of the mechanical work acts as activation energy and is-*via* mechano-electro-optical transformation mechanism - emitted as optical radiation.

4.1.2.3. Emission of Electrons

It is believed that the electrons were due to exothermic process occurring at the surface. It is assumed that the machining process acts as activation process for the emission of electrons (EE). However, owing to the complexity of the sequence of events involved in the whole process, a complete theory of EE effect is not yet available. Recent experimental work on the EE of clean annealed magnesium single crystal surface verifies that electron emission occurs already from a strain free surface simply upon adsorption of oxygen. On the other hand, measurements of EE rates during friction experiments on aluminum surface show that there are close connections between the emission rate of the electrons and the composition of the topmost surface layers and the friction coefficients. Electron can play an important role in the chemical-mechanical process occurring in the machining technique.

In summary, analysis of the effects above all shows that there are some mechanical behavior induced emission processes that must be include in the whole picture of the dissipation of energy in MNT. These processes are physically very interesting and may be very important for the chemical-mechanical process occurring in the system. The basic reason for energy dissipation is the temperature of some local area is appreciably higher than their surroundings, resulting in the transfer of the thermal power, by means of conduction or radiation or the above methods. The transfer of thermal power down a temperature gradient is a unique work process in that work involved appears as thermal power. The net result is that in thermal conduction or radiation, the total thermal-power-flow rate remains constant

$$\dot{Q}_j - \dot{S}_r T_r = \dot{S}_j T_j \qquad (5.37)$$

Here $\dot{S}_r$ is the total flow rate of entropy across a surface at a distance r from a source of entropy, T_r is absolute temperature surrounding a source of entropy, S_j is a source of entropy, T_j is the absolute temperature of S_j, $\dot{Q}_j$ is the associated heat-flow-generation rate. Thermal power can be stored, in which case the location of the storage increases in temperature and the temperature drops when stored thermal power is give up. In general, energy dissipation in MNT is related with the temperature changes affect the sliding friction at tool workpiece interface and materials removal process.

4.2. Microscopic View

Credible microscopic mechanisms for energy dissipation in mechanical processes have been sought for at least the past several centuries. Several models have been derived and proposed to explain the mechanisms of frictional motion from the gross surface ratcheting action offered by Coulomb to the apparently more sophisticated molecular models of the present day. All of these models essentially describe the relative motion of entities between, or over, physical barriers, but few have truly addressed the nature of the energy dissipation mechanism itself. The reason for people know little about mechanical dynamic behavior induced energy dissipation process is that people cannot see what is taking place at the interface. However, as the device such as atomic force microscope have been used to perform friction measurements, characterize contact conditions and even describe the "worn surface". Following these and other experimental developments, machining simulation at the atomic level-particularly molecular dynamics simulations-has brought scientists a step closer to "seeing" what takes place during sliding contact. With these investigations have come some answers and new questions about the modes and mechanisms of energy dissipation at the sliding interface.

MNT can now be studied at the atomic scale, thanks to developments in the past decade of a variety of experimental techniques. The opening of experimental studies of machining at the nanometer and nanosecond scale has attracted theorists equipped to model physical and chemical processes at these scales. Surface scientists are now using sophisticated solid-state potentials to calculate mechanical interactions between surfaces. Force calculations are performed either analytically or by molecular dynamics (MD) simulations. MD simulation affords the added opportunity of using video animation to study MNT processes. For the first time, scientist can "see what is happening" at the otherwise buried sliding interface. They can see atomic and molecular excitation modes (vibrations, bending and rotations), electron-hole excitations, density waves and molecular inter-digitation, to name a few. Once the modes are identified, physicists and chemists can address perhaps the most fundamental but least understood aspect of MNT: energy dissipation processes. While it has been known for centuries that most energy dissipates as heat, neither the macroscopic nor microscopic mechanisms of energy dissipation have been fully explained. Clearly, if friction processes in the MNT generate heat, they are irreversible and cannot be treated by classical thermodynamics. If, however, each step in a sliding interaction is executed with infinite slowness and with the two couples always in equilibrium (never an unbalanced force), then the process may be considered reversible. In such a quasi-static process, sliding can be achieved with zero friction. Real systems can approximate such reversible, adiabatic processes so long as the rate at which each step is taken is much slower than the relaxation time of the system. However, any instability in the mechanical system that leaves unbalanced forces will result in an irreversible process in which energy is lost (or, more precisely, unrecoverable). In a mechanical system, where force, F, acts over a distance, r, the fraction of energy, U, lost over a cycle is given by

$$\Delta U = \oint F \cdot dr \qquad\qquad (5.38)$$

In atomic sliding calculation, a cycle can be a translation across some periodic distance of the lattice, *e.g.*, one atomic spacing. In an experiment, one cycle can be a single pass over a surface, including the making and breaking of the contact.

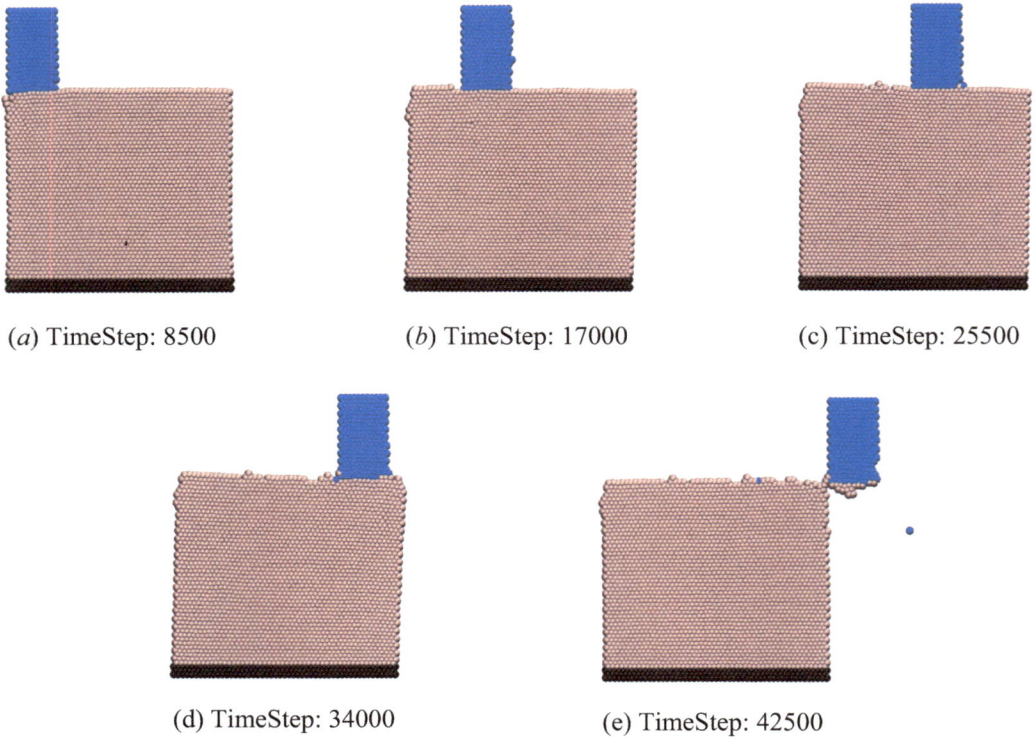

(*a*) TimeStep: 8500   (*b*) TimeStep: 17000   (c) TimeStep: 25500

(d) TimeStep: 34000   (e) TimeStep: 42500

Figure 3: MD simulation results (sliding velocity is 50m/s).

Theoretically speaking, friction in the MNT includes two basic processes, namely, mass transfer and heat transfer. Mass transfer in friction mainly means the wear of materials while the heat transfer is related with energy generation and dissipation. Presently, people know little about the energy dissipation process at the atomic scale and cannot give a convincing explanation about this process because researchers cannot see what is taking place at the interface during sliding. Figs. (**3-4**) gives the MD simulation of sliding and the corresponding dynamic interaction potential energy ($E = \sum_{j=1}^{m} \sum_{i=1}^{n} u_{ij}$, u_{ij} is the interaction potential between the atom of the upper part ($j=1,...m$) and the atom of the lower part ($i=1,...n$)) variation tendency. First of all, the negative interaction energy between the two surfaces remains constant which means no energy dissipation is induced in the initial stage. After that, the interaction energy quickly decreased and after that increased again at the end of the sliding process which also resulted in the variation of the system energy. It can be found that wear of materials is generated

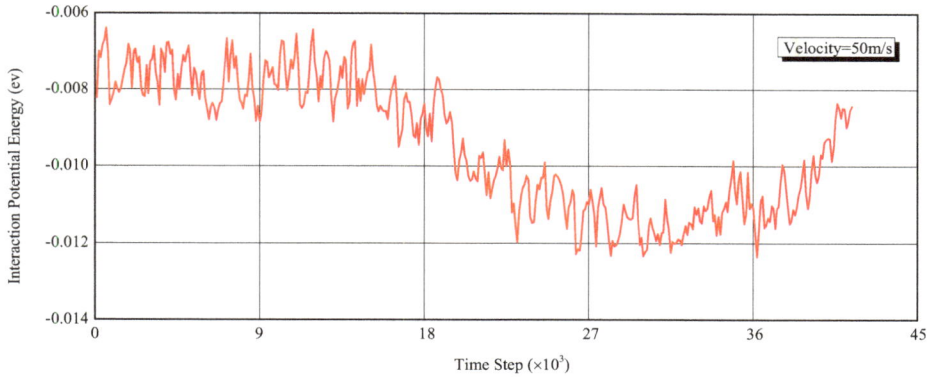

Figure 4: Variation tendency of interaction energy at interface.

in the second part of the sliding friction which changes the surface atom structure and the interaction between friction pair. The decreasing of interaction potential means that the two surfaces come into fusion but this process is gradually weakened with the continually relative sliding between friction pair and finally leads to the fracture of the nanoscale junction and recovery of the interaction potential energy. This mechanical nonlinear dynamic behavior is one of the key factors which influence the energy dissipation of the system. As the above process does not occur under equilibrium conditions, this sliding process is irreversible and the energy lost cannot be reused to assist the sliding process. Stated in more physical terms, the strain energy put into stretching materials surface atoms is not recovered totally, instead, it ought to be converted into molecular vibrational motion which dissipate into the substrate as heat energy. One of the newest issues that tribology must deal with is the concept of matching time and lengths scales in friction studies. As we saw above, energy dissipation hence friction is intimately linked to time and length scales. Moreover, the atomic/molecular modes of interfacial interactions operate at time and length scales far shorter than traditional measurements. The power of microscopic modeling enables us to gain new insights into microscopic essence of phenomenon in MNT.

5. MULTISCALE AND FRACTAL PROPERTY OF MACHINED SURFACE

A surface can be defined as a border between a machined workpiece and its environment. The term surface integrity describes the state and attributes of a machined surface and its relationship to functional performance. Engineering surface is created by a large variety of manufacturing processes such as turning, milling, rolling and lapping. Furthermore, surface alterations may include other methods such as metallurgical, chemical and so on. Functional realization of many precision part depend on materials bulk properties while for a large group of phenomenon these properties are surface properties. Irregular microgeometry includes special surface structure (pit, groove, stepper *etc.*) and surface texture is observed in the machined even undergoing precision or ultraprecision manufacturing process. In general, description of surface physical property can be divided into two aspects, the external topography of surfaces (surface finish) and the microstructure, mechanical properties and residual stresses of the internal subsurface layer. Materials surface determine many technical functions such as anti-wear, anti-fatigue, corrosion, seal, tribology behavior, heat conduction and so on. The surface morphology has a great impact upon the function of parts and its using life. For example, small irregularity of optical surface may lead to the light scattering will degrade the image quality. Investigation of surface morphology is not only of great importance about analyzing the defect generation mechanism and strength technique control but also helpful about improving product quality and function. Presently, there are three statistical characteristics used to describe the structure of machined surface topography, texture, waviness and roughness.

Practically, the deviation of geometry from designed requirement is happened for any surface of mechanical part even it is finished by MNT. These errors embody the geometrical property of materials surface which directly influence the performance of the part. The three kinds of geometrical shape error which makes up for the total multiscale property of machined surface are illustrated as follows and illustrated in Fig. (**5**).

5.1. Macro-Scale Geometry Error (Errors of Form)

This kind of error is induced by the low precision level of machine tool which is a series of continuous and non-repeated surface geometry deviation from nominal geometry, which is named as flatness error, roundness error and cylindricity error and so on. This kind of error has a great effect upon function of mechanical part which can lead to the uneven clearance in clearance fit and the excessively wear of local cylindrical surface. Macroscopic geometry error would decrease the actual supporting area and increases the surface deformation thus intensifying the wear of materials surface.

5.2. Meso-Scale Geometry Error (Waviness)

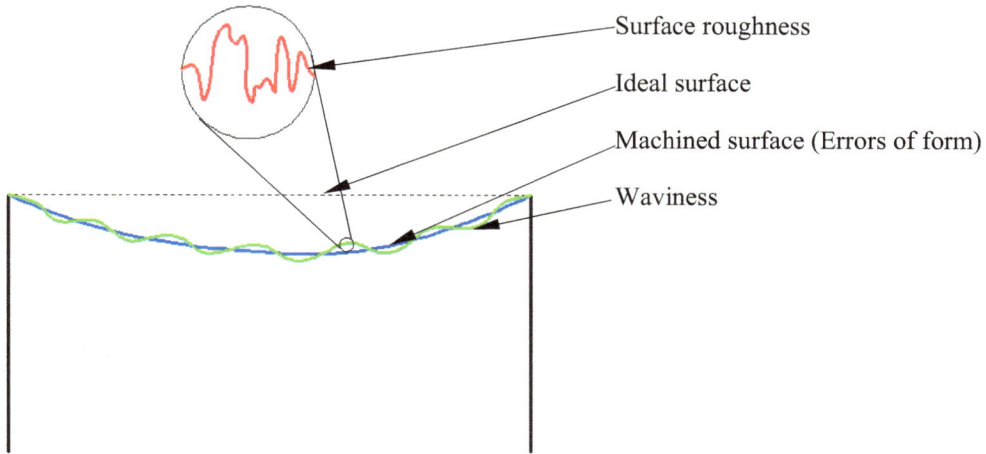

Figure 5: Multiscale surface morphology.

This kind of error is induced by the vibration induced by the machining system composed of machine tool-cutting tool-workpiece and is occurred periodically in the materials surface. This kind of error is usually specialized by means of waviness while the pitch of the waviness is about $1 \sim 10$ mm. The waviness in the materials surface can decrease the effective supporting area and thus intensifying the wear of materials surface.

5.3. Mico-Scale Geometry Error (Roughness)

This kind of error is generated by tool path in metal cutting or plastic deformation in the materials surface which is named as roughness. There are no obvious periodicity about this kind of error while the pitch of the waviness is about $2 \sim 800$ μm. The larger surface roughness can lead to the smaller real contact area at the interface and a much more intense wear. Larger friction resistance will be induced in order to overcome the shear deformation of rough peak between friction pair with larger surface roughness.

Many machined surface topographies are multiscale and random which strongly influence a lot of physical phenomenon related with scientific and engineering applications. Additional levels of roughness details keep appearing when the multiscale topography is successively magnified. Furthermore, the magnified surface structure appears very similar to the original surface which suggests that roughness is statistically self-similar or self-affinity. The multiscale nature of roughness warrants a scale independent characterization. It is necessary to identify advantages and disadvantages of classical surface characterization method before a new method is discussed. One of the unique features of rough surfaces is that it has a multiscale structure for which the range of roughness can span from large length scales right down to atomic scales. Presently, most of the surface characterization method use several scale dependent parameters to represent the surface geometry. In reality, physical phenomenon is influenced by all scales of roughness from the largest scale down to the length scale of the phenomenon. However, if the largest scale of the surface is small than the length scale of phenomena then the surface seems to be smooth. As the length scales of different phenomena are not likely to be the same, it is therefore logical to characterize the roughness structure in such a way that the roughness structure is known at any scale. Another unique feature of rough surface is that unless deliberately textured or machined, the roughness usually appears to be disordered and irregular.

With gradually improvement of precision of machining and measurement, some novel concepts such as sub-roughness and atomic roughness is put forward for the requirement of nanometer level machined surface morphology. The classical method on identification and evaluation of surface morphology is mainly focused on measuring and characterizing of surface roughness. The surface roughness justifies the microscale geometry feature which is difficult to be controlled in the machining and also difficult to be measured and evaluated. Therefore, the study of surface roughness is the core content of surface geometry research. It is difficult to fully characterize surface morphology because of the irregularity, uncertainty and complexity of machined surface. With the deepening of the study of surface morphology, various parameters are continuously added in the existing surface characterization system, part of them are contradictory while part of them have the same evaluation but with different name or expression. A phenomenon named as "parameter rash" is induced by the increasing of parameters which is generated by two basic situations: firstly, the realization that the existing parameters are largely unhelpful in describing the way a surface would perform against a given "functional application"; secondly, high performance computer aid researchers about digitizing the formerly analogue results and thus proliferating the new parameters. This achievement has encouraged and enabled equipment manufactures to introduce PC-based commercial systems which can provide both of surface measurement and the corresponding analysis. It is always taking hours to implement a new parameter idea while their inclusion added little to the general understanding of surface morphology. The successively growth of the famed "parameter rash" confuse both of designers and engineers involved, and lead to the controversy, argument and doubt in communication of the subject.

On the other hand, the main object of MNT is to minimize the damage created due to the removal of materials at high materials removal rates. A substantial increase in the specific energy required with a decrease in chip size during machining. It is believed this is due to the fact that all metals contain defects such as grain boundaries, missing and impurity atoms, *etc.*, and when the size of the material removed decreases the probability of encountering a stress-reducing defect decreases. Since the shear stress and strain in metal cutting is unusually high, discontinuous microcracks usually form on the primary shear plane. If the material is very brittle, or the compressive stress on the shear plane is relatively low, microcracks will grow into larger cracks giving rise to discontinuous chip formation. When discontinuous microcracks form on the shear plane they will weld and reform as strain proceeds, thus joining the transport of dislocations in accounting for the total slip of the shear plane. A novel surface generation process with different control mechanism is induced in MNT which also need some new method to characterize the micro & nanomachined surface.

5.4. Fractal Geometry

Euclid and his students introduced the concept of space dimension which can take positive integer values equal to the number of independent directions. A smooth line has dimension 1, a smooth surface has dimension 2 and our space (seen at large scales) has dimension 3. In the nineteen century, people started to study the generalization of dimensions to fractional values. The concept of non-integer dimensions was fully developed in the first half of the twentieth century. To capture this novel concept, the word "fractal" was coined by Mandelbrot [6], from the Latin root fractus to capture the rough, broken and irregular characteristics of the objects he intended to describe. In fact, this roughness can be present at all scales, which distinguishes fractals from Euclidean shapes. Mandelbrot worked actively to demonstrate that this concept is not just a mathematical curiosity but has strong relevance to the real world. The remarkable fact is that this generalization, from integer dimensions to fractional dimensions, has a profound and intuitive interpretation: non-integer dimensions describe irregular sets consisting of parts similar to the whole. This generalization of the notion of a dimension from integers to real numbers reflects the conceptual jump from translational invariance to continuous scale invariance. According to Mandelbrot, a fractal is a rough or fragmented geometric shape that can be subdivided into parts, each of which is (at least approximately) a reduced-size copy of the whole. Mathematically, a fractal is a set of points whose fractal dimension exceeds its topological dimension.

Experience has shown that it is often difficult to be precise when defining fractals. The definition based solely on dimension is too narrow and it is better to view a fractal set as possessing a fine structure, too much irregularity to be described in traditional geometric language, both locally and globally, some form of self-similarity, perhaps approximate or statistical, a "fractal dimension" (somehow defined) which is greater than its topological dimension, and a simple definition, usually recursive. According to fractal geometry, a complex phenomenon seems to be out of order may be scale independent or self-similarity.

The study justifies that surface is not simply described with the integer (1, 2 or 3) dimension, but can be described with a fractional number between 1 and 2 for a profile and 2 and 3 for the surface. This technique provides the possibility to describe the intrinsic properties of surfaces with some fractal parameters such as fractal dimension and topthesy without the influence of measurement scale or sampling interval. It has been proved by many authors that engineering surfaces are fractal. They are statistically "self-affine", *i.e.* the scales in the vertical and lateral directions are different rather than "self-similar" which means that a self-similar random fractal has the same statistics in relation to the scale of observation for all scales in all directions.

5.5. Self-Similarity and Self-Affinity

There are two terms that are relevant in terms of scale. One is 'self-similarity' which refers to the property that it has the same statistics regardless of scale. In terms of roughness parameters therefore self-similarity implies that any parameter should be independent of the scale of sampling. The second, a 'self-affine' fractal, is only self-similar when scaled in one direction. Single-valued fractal functions must be self-affine because a small feature can only occur on a larger feature if its slope is larger. On this basis a surface profile should preferably be considered to be single-valued because surface slopes are steeper for finer sampling. The difference between the self-similar property and the self-affine property is that self-similar fractals require only one parameter to define them - D, the fractal dimension - whereas self-affine surfaces require an additional one that relates to the scale being viewed. In roughness terms it is necessary to specify, in addition to D, the way in which the ratio of vertical to horizontal magnification has to be changed in order to preserve self-similarity. The name given to this other parameter is called "topothesy".

A self-affine object looks the same after an affine transformation: If a small piece of the object is stretched with different ratios in different directions, then the enlarged object recovers (statistically recovers) the original object. Theoretically speaking, a rough surface can be described by a sing-valued self-affine function:

$$h(x_1, x_2, \cdots, x_n) = \varepsilon_1^{\alpha_1} \varepsilon_2^{\alpha_2} \cdots \varepsilon_n^{\alpha_n} h(\varepsilon_1 x_1, \varepsilon_2 x_2, \cdots, \varepsilon_n x_n) \qquad \textbf{(5.39)}$$

here h is the surface height, α_i is called roughness exponent. Typically, there is only one characteristic roughness exponent and the equation (5.39) has another simpler form,

$$h(x) = \varepsilon^\alpha h(\varepsilon x) \qquad \textbf{(5.40)}$$

For example, for a single variable x, equation (5.39) demonstrate the fact that the function h is invariant under the following rescaling: shrink the variable x along the x-axis by a factor of $1/\varepsilon$, rescale the value of the function by a different factor $\varepsilon^{-\alpha}$. Therefore, a self-affine surface is a class of fractal objects that can be described by a roughness exponent which is related to the fractal dimension of the surface and the height-height correlation function of such a surface has the form:

$$H(r) = 2\omega^2 f\left(\frac{r}{\xi}\right) \tag{5.41}$$

here $f(x)$ is a scaling function which having the following properties:

$$f(x) = \begin{cases} x^{2\alpha}; & x \leq 1 \\ 1 & ; & x > 1 \end{cases} \tag{5.42}$$

here α is called the roughness exponent which describes how wiggly the surface is. The asymptotic behavior of the height-height correlation function can be written as

$$R(r) = f\left(\frac{r}{\xi}\right) \tag{5.43}$$

The height-height correlation function can also be described as phenomenological function:

$$H(r) = 2\omega^2 \left\{ 1 - \exp\left[-\left(\frac{r}{\xi}\right)^2 \right] \right\} \tag{5.44}$$

This form reflects the asymptotic behavior of the scaling function $f(x)$ in a simple and straightforward way.

5.6. Fractal Dimension Calculation (Fig. 7)

An object, say two-dimensional, such as a square area in the plane, can be divided into N self-similar parts, each of which is scaled down by a factor $r = \dfrac{1}{\sqrt{N}}$. For line segment, $N=r^{-1}$, and solid cube, $N=r^{-3}$. A D-dimensional self-similar object can be divided into N smaller copies of itself each of which is scaled down by a factor r:

$$N = \left(\frac{1}{r}\right)^{D} \tag{5.45}$$

If we change r to br, a self-similarity relationship can be obtained as follows:

$$N(br)=b^{-D}N(r) \tag{5.46}$$

The the fractal dimension is given by

$$D = \frac{\ln N}{\ln \dfrac{1}{r}} \tag{5.47}$$

Eq. (5.45) can be changed as

$$\frac{1}{N} = r^{D} = v \tag{5.48}$$

It can be deduced that if the generalized volume of a geometrical object is enlarged (or reduced) by a factor v with the enlargement (of reduction) of a factor r in linear dimension; then, the fractal dimension is given by

$$D = \frac{\ln v}{\ln r} \tag{5.49}$$

5.6.1. Hausdorff-Besicovitch Dimension

Now, we turn back to the tentative definition of fractals given by Mandelbrot. This definition requires previous definitions of *Hausdorff-Besicovitch* dimension and topological dimension. The *Hausdorff* dimension D of a subset S of Euclidean space arises from asking "how big is S" for very general sets. The answer comes from counting the number of open balls needed to cover the set S. For each $r > 0$, let $N(r)$ denote the smallest number of open balls of radius r needed to cover S. One can show that the limit

$$D = \lim_{r \to 0} \frac{-\ln N(r)}{\ln r} \tag{5.50}$$

exists. The value of D is called the Hausdorff dimension of S. Since $\ln r \to -\infty$ as $r \to 0$, the negative sign is needed in order that D should be positive.

Eq. (5.50) is equivalent to an approximate power-law relationship

$$N(r) \approx const. \times r^{-D} \tag{5.51}$$

Assume S can be decomposed into n rescaled copies of itself, each contracted by a linear factor k, then we can get

$$D = \frac{\ln n}{\ln k} \tag{5.52}$$

here $k = r^{-1}$.

If we suppose that S can be covered by $N(r^0)$ open balls of radius r^0. Each reduced copy can be covered by n "rescaled" open balls of radius $\frac{r_0}{k}$. Then, S can be covered by $nN(r^0)$ open balls of radius $\frac{r_0}{k}$. At least approximately,

$$N(\frac{r_0}{k}) = nN(r_0) \tag{5.53}$$

The iteration of the above process m times gives

$$N\left(\frac{r_0}{k^m}\right) = n^m N(r_0) \quad (m=0,1,2,3,\cdots) \tag{5.54}$$

Eq. (5.54) means that

$$\lim_{m \to \infty}\left[\frac{-\ln N\left(\dfrac{r_0}{k^m}\right)}{\ln \dfrac{r_0}{k^m}}\right] = \lim_{m \to \infty}\frac{\left[m \ln n + \ln N(r_0)\right]}{\left(m \ln k - \ln r_0\right)}$$

$$= \ln n / \ln k \qquad\qquad\qquad\text{(5.55)}$$

$$= D$$

Therefore, the scaling dimension is equivalent to the *Hausdorff* dimension. The condition that the limit of Eq. (5.50) exists is equivalent to finding a real value D such that for the set S, the d-measure is infinite if $d < D$ and vanishes if $d > D$. D is called the *Hausdorff* dimension of S, and is also called the *Hausdorff-Besicovitch* dimension when non-integer values of D are included.

5.6.2. W-M Fractal Function

The modified *Weierstrass-Mandelbort* (*W-M*) function for a surface is continuous but not differentiable, and exhibit self-affinity. It is suitable for characterizing the surface, and can be described as:

$$Z(x) = A^{(D-1)} \sum_{n=n_1}^{\infty} \frac{\cos 2\pi \gamma^n x}{\gamma^{(2-D)n}} \quad 1<D<2,\ \gamma > 1 \qquad\qquad\text{(5.56)}$$

where $Z(x)$ is the surface profile, D is the fractal dimension, A is the characteristic length scale of a surface, n is the discrete frequency spectrum of the surface roughness, n_1 corresponds to the low cutoff frequency of the profile. It can be seen from Eq. (5.56) that the shape of a surface profile can be determined by D and A uniquely, and their values are independent of sampling length and resolution of the measuring instrument for a particular surface.

5.6.3. Power Spectrum Method

Research shows that many engineering surfaces exhibit fractal features. And the spectrum of the surfaces follows power laws of the form:

$$P(\omega) = \omega^{-k} \qquad\qquad\qquad\qquad\text{(5.57)}$$

Where $P(\omega)$ is the power of the spectrum, ω is the frequency of the surface roughness of the spectrum, k is a coefficient, and fractal dimension $D = (5 - k)/2$. If $P(\omega)$ is plotted as a function of ω on a log-log graph, we can get the slope $-k$ of a fitting line, and then we can get the fractal dimension D.

5.6.4. Wavelet Transform Method

Wavelet transform method estimates fractal dimension using wavelet coefficients, considers all points on the surface profile curve as a discrete sequence $z(x)$. Its mathematical description is as follow:

$$\frac{\sum_{n=1}^{2^k}\left|d_{k,n}\right|}{2^k} = E2^{-k(5/2-D)} \tag{5.58}$$

where E is a constant, k is an integer, D is the fractal dimension, $d_{k,n}$ is the wavelet

coefficients. If $\dfrac{\sum_{n=1}^{2^k}\left|d_{k,n}\right|}{2^k}$ is denoted by f, Eq. (5.58) can be defined as:

$$\log_2 f = \log_2 E - k\left(\frac{5}{2} - D\right) \tag{5.59}$$

Hence we can calculate the parameters E and D from the regression line of a sufficient number of points.

5.6.5. Variation Method

Consider the surface profile curve as a continuous function $z(x)$, and for all x in $[0,1]$. The ε oscillation of the function z in x is given by:

$$v(x,\varepsilon) = \sup_{x'\in R(x)} z(x') - \inf_{x'\in R(x)} z(x') \tag{5.60}$$

where $R(x) = \{c : |x - c| < \varepsilon, c \in [0,1]\}$, and the ε variation of the function z is given by

$$V(\varepsilon, z) = \int_0^1 v(x,\varepsilon)dx \tag{5.61}$$

Since z is continuous, the ε variation of z tends to 0 as ε tends toward 0.

Suppose that the digitized profile data are $z(n/N)$, and the total sampling points are $N + 1$, $n = 0, 1,., N$. Given a list of integers in increasing order k_i, $i = 1, 2,., i_{max}$. The ε variation of z after separation can be described by

$$V(\varepsilon_i, z) = \int_0^1 v(x, k_i / N)dx \tag{5.62}$$

with $\varepsilon_i = \dfrac{k_i}{N}$

$u_i(n)$ and $b_i(n)$ are defined as follow:

$$u_i(n) = \max\{z(j / N); \ j \in [n - k_i, n + k_i]\}$$

$$b_i(n) = \min\{z(j / N); \ j \in [n - k_i, n + k_i]\} \tag{5.63}$$

Then

$$V(\varepsilon_i, z) \approx \frac{1}{N + 1} \sum_{n=0}^{N} u_i(n) - b_i(n) \tag{5.64}$$

Different $\dfrac{1}{\varepsilon_i}$ and the corresponding $\dfrac{1}{\varepsilon_i^2} V(\varepsilon_i, z)$ can be plotted on log-log graph and then we can get the slope, the fractal dimension.

5.6.6. Difference-Average Law Method

Difference-average law method is easy to implement by numerical algorithm. And it is very suitable for processing the discrete data obtained by profiler. Its mathematical description is as follow:

$$b(m) = \sqrt{\frac{2}{\pi}} \sigma [a(m)]^{2-D} \tag{5.65}$$

with $b(m) = \dfrac{1}{n - m} \sum_{k=1}^{n-m} |z_{k+m} - z_k|$ and $a(m) = \dfrac{m \Delta x_0}{1mm}$, where D is the fractal dimension, m is an integer, σ is a coefficient, Δx_0 is the lateral distance between two subsequent measuring points. Different $a(m)$ and the corresponding $b(m)$ can be plotted on a log-log graph, we can get the slope $2 - D$ of a fitting line, and then we can get the fractal dimension D.

5.7. Fractal Characterization of Machined Surface

Standard W-M fractal function curves whose fractal dimension are 1.2, 1.5 and 1.8 are shown in Fig. (**6**). The theoretical and estimated fractal dimension values are mention in Table **1**. It is apparent that the real surface profile is similar to the curve of $D = 1.5$. Moreover, variation method and power spectrum method

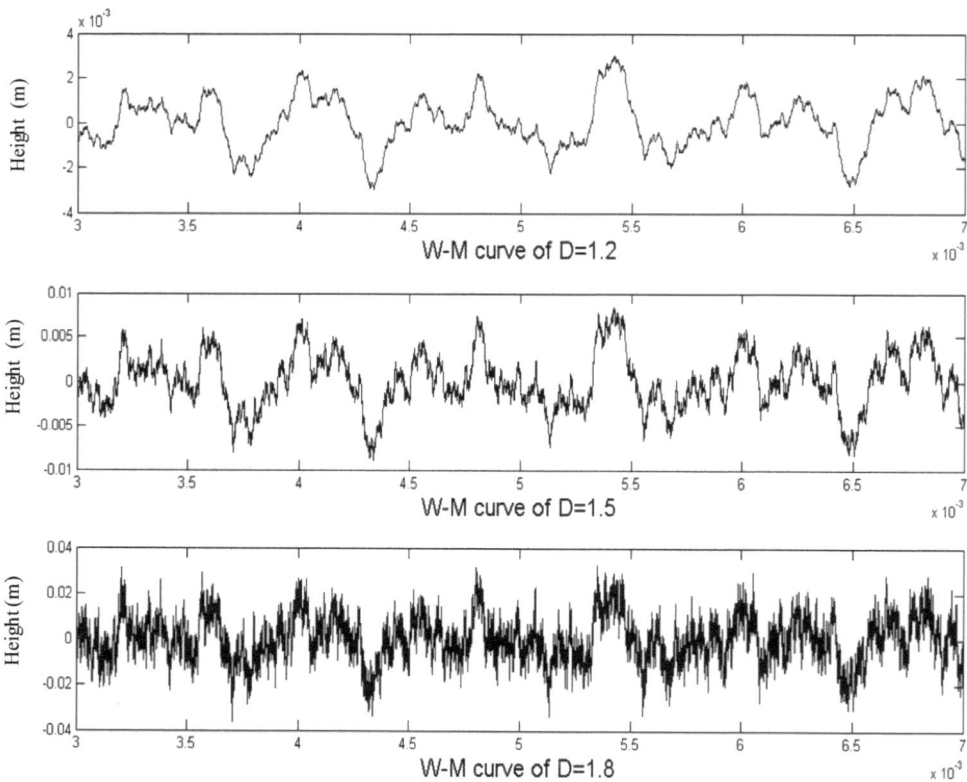

Figure 6: Fractal curves based upon standard W-M function.

Table 1: Estimated and theoretical fractal dimension.

Theoretical fractal dimension	1.2	1.3	1.4	1.5	1.6	1.7	1.8
Estimated fractal dimension	1.2377	1.3090	1.3890	1.4734	1.5597	1.6393	1.7163

provide the similar calculation results, but results of difference-average law method and power spectrum method are quite different. Therefore, difference average law method is of low precision when it is applied to the engineering ceramics surface. We found that difference-average law method and wavelet transform method have high precisions when they calculate the standard W-M fractal function curve, however, they have low precisions when applied to the engineering ceramics surface. The possible reason for this phenomenon is that: engineering ceramics surface is easy to emerge cracks, collapses, and other features during the machining process, and is not rigid fractal, while variation method is more suitable for such surface.

In traditional roughness parameters, arithmetical mean deviation of the assessed profile R_a is the most common one. Based on the statistics of a large number of measurement data, we found that there is no one to one correspondence between arithmetical mean deviation of the assessed profile R_a and fractal dimension D. The fractal dimension of surfaces with the same material and R_a value may be significantly different. Furthermore, the texture of engineering ceramics usually mixes with cracks, collapses, and has poor continuity. It is significantly different from texture of metal surface. Based on the statistics of a large number of measurement data, we found that: In general, the texture is fine and dense, when the fractal dimension is high; the texture is coarse and spars, when the fractal dimension is low. With the zirconia specimens for $R_a = 0.4\mu m$ as example, the fractal dimensions of two specimens are 1.43, 1.53, respectively.

Fractal analysis has been increasingly applied to many aspects of surface investigation such as surface modeling, surface characterization, surface contact and wear analysis. Many fractal functions such as *Weierstrass-Mandelbrot*, fractal Brownian motion and Takagi functions proposed since the original work of Mandelbrot to model fractal surface, and many methods such as power spectrum, variation method have been adopted to calculate the fractal dimensions of surface. It was reported that a self-affine engineering surface can be completely described by just two fractal parameters, *i.e.* the fractal dimension and the topothesy. A change of surface topography due to wear and other process may give rise to a change of fractal dimension and/or the topthesy. Thus the two fractal parameters may avoid parameter rash and scaling effects on surface topography characterization.

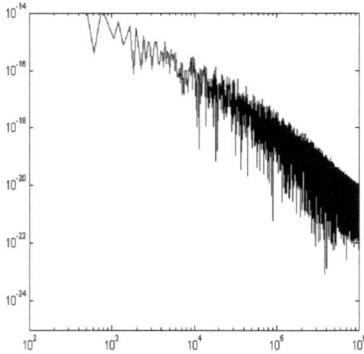

(a) Power spectrum method

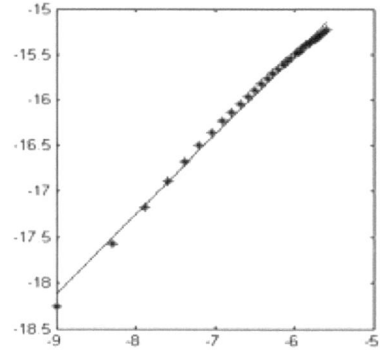

(b) Difference-average law method

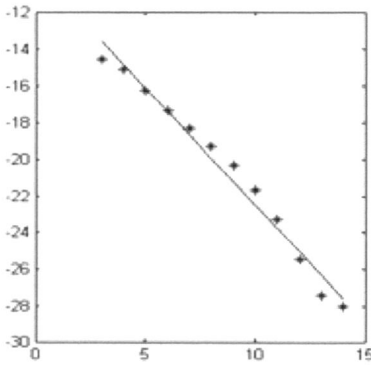

(c) Wavelet transform method

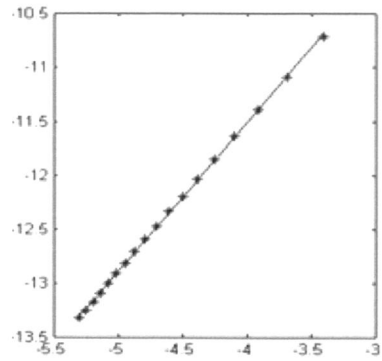

(d) Variation method

Figure 7: Fitting results of different fractal computation method.

Currently, surface topography characterization methods of engineering ceramics are similar to the metal, which mainly use a series of statistical parameters. In traditional statistical approach, we need to anticipate the range of roughness, and choose the appropriate sampling length and filter (cut-off) value. The measurement results will be obviously different if we choose different sampling length and filter value for the same profile. Therefore, the traditional surface topography characterization method still has some shortcomings. Fractal method is introduced to characterize the engineering ceramics ground surface. This method can provide parameters which are independent of sampling length and resolution of the measuring instrument, and can reflect intrinsic property of surface. After thoroughly investigation, the following conclusions are drawn:

1. The variation method is suitable for extracting fractal dimension of engineering ceramics ground surface.

2. There is no one to one correspondence between arithmetical mean deviation of the assessed profile R_a and fractal dimension D. The fractal dimension of surfaces with the same material and R_a value may be significantly different. In general, the texture in engineering ceramics surface is fine and dense, when the fractal dimension is high; the texture is coarse and sparse, when the fractal dimension is low.

3. With the increasing of fractal dimension, the material ratio of the profile (*pmr*) of various materials in general also tends to increase. Namely with the higher value of the fractal dimensions, they can indicate the larger contact area and the better wear resistance for the same material. Meanwhile, the size order of fractal dimension is steel, zirconia, silicon nitride and alumina. It is the same as the order of fracture toughness of four materials.

There is still a lot of work to do for the application of fractal method in engineering ceramics surface, for example, the calculation method of the characteristic length scale A and the relationship between fractal dimension and various surface functions. Therefore, characterization of engineering ceramics surface topography still needs to depend on traditional statistical approach currently. But fractal approach can reflect the differences among various materials in surface topography and material property, the introduction of the fractal approach makes up for the deficiency of traditional statistical approach to some extension. Fractal parameters lack straightforward physical meanings, and have not been directly linked with any functional applications. The variation of fractal parameters in terms of surfaces and characterization procedures has not been fully investigated yet. Furthermore, there is little suggestion as to whether fractal analysis might only be used as a complementary method to characterize engineering surfaces or whether it can be used to substitute the classical techniques. When deepening the research, characterization methods of engineering ceramics surface topography will be increasingly perfect.

REFERENCES

[1]   X.S. Han, "A study on the mechanism of ultraprecision manufacturing and its simulation", M.S. thesis, Tianjin University, Tianjin, China, 2002.

[2]   N.P. Suh, "Complexity in engineering", Annl. CIRP, vol. 54, pp. 581-598, April 2005

[3]   W. K. Liu, E. G. Korpov, H. S. Park, Nano Mechanics and Materials. New York: John Wiley & Sons, 2006.

[4]   B.J. Kostetskii, P.V. Nazarenko, "Influence of changes of dislocation structure on the relation between friction and normal pressure", Soviet Phys., vol. 9, pp. 1011-1017, March 1965

[5]   D.M. Tolstoi, "Significance of the normal degree of freedom and natural normal vibrations in contact friction", Wear, vol. 10, pp. 199-204, August 1967.

[6]   B.B. Mandelbrot, *The fractal geometry of nature*. New York: W H freeman, 1982.

Subject Index